{ AELIAN'S }

On the Nature of Animals

Gregory McNamee

TRINITY UNIVERSITY PRESS
SAN ANTONIO

Published by Trinity University Press
San Antonio, Texas 78212

Book design by Anne Richmond Boston

Trinity University Press strives to produce its books using methods and materials
in an environmentally sensitive manner. We favor working with manufacturers
that practice sustainable management of all natural resources, produce paper using
recycled stock, and manage forests with the best possible practices for people, bio-
diversity, and sustainability. The press is a member of the Green Press Initiative, a
nonprofit program dedicated to supporting publishers in their efforts to reduce their
impacts on endangered forests, climate change, and forest dependent communities.

The paper used in this publication meets the minimum requirements of the
American National Standard for Information Sciences—Permanence of Paper for
Printed Library Materials, ANSI z39.48-1992.

Library of Congress Cataloging-in-Publication Data

Aelian, 3rd cent.
[De natura animalium. English. Selections]
Aelian's On the nature of animals / [translated and edited by] Gregory McNamee.
p. cm.
Summary: "Selections from Aelian's *De Natura Animalium*, translated and edited by
Gregory McNamee, are a mostly randomly ordered collection of stories that consti-
tute an early encyclopedia of animal behavior, affording insight into what ancient
Romans knew about and thought about animals—and, of particular interest to
modern scholars, about animal minds"—Provided by publisher.
ISBN 978-1-59534-075-7 (pbk. : alk. paper)
1. Zoology—Pre-Linnean works. I. McNamee, Gregory. II. Title.
III. Title: On the nature of animals.
PA3821.E5 2011
590—DC22   2010053843
Printed in China

{ INTRODUCTION }

The octopus is greedy, sneaky, and voracious, and it will eat anything.

The owl is a wily bird, as crafty as a sorceress. If you were to capture one, it would beguile and bewitch you so that you would keep it as a pet, even allow it to sit on your shoulders as if it were some sort of good-luck charm.

If a horse should step on a wolf's footprint, then it would go numb.

When cranes squawk, rain is on the way.

If you want to start an argument at a dinner party, you can immediately do so by dropping a stone that a dog has gnawed into the wine. This will whip your guests into a fury.

We owe these observations on the ways of the animals of land, sea, and air to an encyclopedist, writer, collector, and moralist named Claudius Aelianus. Aelian, as we call him, was born sometime between 165 and 170 CE in the hill town of Praeneste, what is now Palestrina, about twenty-five miles from Rome. We do not know much about his early life, but we can imagine him to have been a bookish and curious boy, the kind who, like Heraclitus, might lie alongside a busy road to study the ways of industrious dung beetles and pester grown-ups to teach him how to read auguries from the flights of birds. He grew up speaking his town's version of Latin, but, somewhat unusually for his generation, though not for Romans of earlier times, he preferred to communicate in Greek, the language in which he wrote his many books.

Trained by a sophist named Pausanias of Caesarea, Aelian was known in his time for a work called *Indictment of the Effeminate*, an attack on the recently deceased emperor Marcus Aurelius Antoninus, who was nasty even by the standards of Imperial Rome. He was also fond of making almanac-like collections, only fragments of which survive, devoted to odd topics such as manifestations of the divine and the workings of the supernatural. His *On the Nature of Animals*, which has survived more or less whole, is just such a book, a mostly randomly ordered collection of stories that he found interesting enough to relate about animals, whether he believed them or not.

Aelian's magnum opus constitutes an early encyclopedia of animal behavior. This edition consists of selections from the work, in my translations. If Aelian's science is sometimes sketchy, the facts often fanciful, and the history sometimes suspect, it is clear enough that he had a grand time assembling the material, which can be said, in the most general terms, to support the notion of a kind of intelligence in nature and that extends human qualities, for good and bad, to animals. Modern scientists carefully avoid such anthropomorphizing, but the ancients did not. Aelian's contemporaries enjoyed his stories, praising him for the clarity of his Greek, even though they also considered it a touch old-fashioned, as if he were writing in the English of Charles Dickens (or, for that matter, Charles Darwin) today. Those stories always had a point, and they would have resonated with his audiences, who were able to supply many stories of their own to reinforce his.

Aelian, for instance, need only allude to the legendary harpist Arion, who once was waylaid by pirates and tossed overboard. As the story has it, a passing dolphin rescued him and took Arion to the southern Italian port of Taranto, where he met the arriving pirate ship. The authorities were not amused and hanged the pirates; the dolphin remains Taranto's talisman today. In his *History of Greek Literature* (1957), Albin Lesky observes that the story of Arion fuses two great themes. The first is the one Aelian treats all along through the pages of *On the Nature of Animals*—namely, that the gods work their ways through animal actors as much as human ones. In this instance, the dolphin is sacred to Apollo, and because Arion is both pious and a good harpist, music-loving Apollo sees to it that Arion is protected. The second is a theme that Aelian would have known and relied on: namely, that Greeks and Romans were always telling stories about how helpful and friendly dolphins were with respect to humans. The story of Arion was an old one when it turned up in the pages of Herodotus, its first known recorder, older by centuries more when Aelian got to it. Still, it remained a good story that pointed to the mysterious ways of divine beings and animals alike, which is just why Aelian put it to use alongside many other good stories that he had heard and read.

Aelian died sometime between 230 and 235 CE, but he was not forgotten; Christian writers in particular found that his stories of animals attested to the ways of a wise and forgiving deity. Here is Gerald of Wales, for instance, writing nearly a millennium after Aelian's time in words that might well have come from Aelian himself: "Indeed the soil of this land is so inimical to poison that, if gardens or any other places of other countries are sprinkled with it, it drives all poisonous reptiles far away." Along with other compendia of natural history, such as those of Pliny and the makers of the Egyptian *Physiologus*, Aelian's example endures in the bestiaries that guided our understanding of animals until very modern times, before the rise of science. Just as the mapmaker for Mungo Park's 1793 African expedition found himself "relying very heavily on Herodotus" in drafting what he hoped was a reliable chart of the continent, so readers learned from Aelian and his successors why it was wise not to step in a wolf's footprint, fall asleep in the presence of hyenas, or be surprised if a dolphin were to fall in love with a person innocently lazing away on the shore.

Aelian knew as much as any person of his time about animals. He knew what his contemporaries knew, and he knew what they would find exotic. *On the Nature of Animals* is thus both a wonderful window onto the beliefs of ordinary people and a testimonial to the transmission of knowledge in the ancient world. It is also a great entertainment to read, as Aelian ponders the ways of the animals and tries to work them out, sometimes successfully, by our lights, and sometimes not:

> Nature has given animals many different voices and languages with which to speak, just as it has done with humans. Scythians have one language, and Hindus another; Ethiopians speak one

tongue, and Sacae another; the languages of Greece and Rome are not the same. And so it is with animals. Each one has a different way of speaking. One roars, another moos, another whinnies, another brays, another bleats; some get by with howling, and others barking, and others roaring. Screaming, whooping, whistling, hooting, twittering, singing—these are just some of the ways in which animals speak.

The octopus is a lustful fish, and it couples until all its strength has vanished, leaving it drained and incapable of swimming or hunting. When this happens, little fish and crabs, particularly the kind called the hermit crab, come down and eat the octopus. They say that this is why octopuses live for only a year and no more. The female octopus is exhausted, too, by giving birth so often.

I take the first quotation to represent a significant effort to understand the diversity of animals as well as their minds, our knowledge of which, all these centuries later, we have scarcely improved. And if the second offers a touch too much moralizing for modern sensibilities, it still makes for acceptable postulating. Unseemly attributions to prurience notwithstanding, it was the product of its time and place and of what Aelian was able to say with confidence about the private lives of creatures in their unseen abodes—creatures that, Aelian would suggest, could be just as strange as humans.

Often we find these entries amusing, and rightly so. Often we find them outlandish, foolish, primitive. Yet I suspect that not so long from now—if there is a not so long from now for us bus-ily habitat-devouring humans—scientists will wonder at our own naïveté and arrogance, at the thought that language, emotion, and even reason are the gifts of humans alone. A mere look out the door should disabuse us of such notions, and the more we learn about the mental worlds of the animals with whom we share the world, the more we realize, or at least should realize, that we are kin of different tribes and stand to learn much from our distant cousins. As Aelian remarks succinctly, "It breaks my heart that dogs are more loyal and kinder than men."

His historical and literary significance notwithstanding, Aelian —the Herodotus of the animals, who took such joy in writing and storytelling—has long brought pleasure to readers interested in good storytelling, as well as to lovers of animals and of the natural world. With these selections from his work, as various and as beautiful as wildflowers in an Italian field, I hope to bring that pleasure to readers in our own time, eighteen-odd centuries after Aelian's own.

{ PROLOGUE }

That humans are wise and just, and take care of their children, and honor their parents appropriately, and feed themselves, and protect themselves from others' schemes, and possess other gifts of nature should not strike anyone as unusual. Humans have the gift of speech, the greatest gift of all, as well as of reason, which is always of use. Humans also know how to revere and worship the gods. It is, however, remarkable when animals, which have none of these gifts or knowledge, should nevertheless exhibit some of the best qualities of humans. It requires a sharp mind and a great deal of learning to discern the characteristics of animals, and many scholars have turned to these questions over the years. For my part, I have gathered everything I could learn on the subject here, and put it all into ordinary speech. It seems to me that the result is noteworthy. If you think so, then I hope these words will be useful to you. If not, give them to your father to keep and study. Not everything pleases everyone, and not everyone wants to study everything. Plenty of other writers have come before me, but that should not disqualify me from praise, if it really is true that this learned book is far ranging and well written enough to deserve attention.

{ BOOK I }

An island in the Adriatic Sea, called Diomedea, is home to great numbers of **SHEARWATERS**, which, it is said, neither harm the barbarians who live there nor come close to them. If, however, a Greek comes ashore, the shearwaters approach and stretch out their wings as if they were hands, welcoming the stranger. They will let a Greek touch them, and they will even fly up and sit on a Greek's lap, as if they had been invited to supper. The shearwaters, the story goes, were once the companions of Diomedes in the war against Troy, and even though they were transformed somehow from humans into birds, they nonetheless retain their Greek nature and love of their compatriots.

The **MULLET** lives in pools and controls its appetites, so that it leads a very moderate life. It never attacks a living creature, but instead is inclined to be peaceable with all other fish. If it comes across a dead fish, the mullet will eat it, but only after giving it a swat with its tail. If the fish does not move, then the mullet considers it fair game, but if it moves, then the mullet backs away.

A **DOG,** it is said, fell in love with Glauce the harpist. Some who tell the story say that it was not a dog but a ram, while still others say that it was a goose. At Soli in Cilicia, a dog fell in love with a boy named Xenophon. And at Sparta, a jackdaw fell in love with a beautiful young man.

It is easy to tell how old a **BEE** is. Those born in the past year gleam and are the color of fresh olive oil, while the older ones look wrinkled. The older ones, of course, are more skilled in the making of honey, and they also have powers of divination, so that they know when rain and cold will arrive. When they sense that rain or cold are indeed on the way, they fly close to their hives and do not venture far abroad. An observant farmer can thus tell when a storm is on the way simply by observing the behavior of the bees, which have less fear of frost than they do of heavy rain or snow. In times of heavy wind, the bees will sometimes pick up a pebble with their legs and use it as ballast, and in this way they are not blown off course.

Some **FISH** are models of good behavior. There is one that swims in the vicinity of Mount Etna that mates for life and does not approach any other fish of the opposite sex. It needs no laws, no customs, no dowry; it just does this naturally. This is a noble law of nature, worthy of honor. But humans, inclined to do what they want to, are not at all ashamed to disobey it.

The **HORNED RAY** is born in mud. It is very small at birth, but it grows to a huge size. Its belly is white; its back, head, and sides are inky black. Its mouth, though, is small, and you cannot see its teeth. It is very long and flat. It eats great quantities of fish, but its favorite food is human flesh. It has little strength, but its size gives it courage. When it sees a man swimming or diving, it rises to the surface, arches its back, and slams down on him with all its might, extending its length over the unfortunate man like a roof and keeping him from rising to breathe. The man dies, and the ray greedily enjoys its feast.

They say that the **JACKAL** is the kindliest disposed of all animals toward humans. Whenever a jackal comes across a human, it will step out of the way, and whenever it sees that some other animal has harmed a human, it comes running to the rescue.

There was once a man named Nicias who became separated from his hunting party and fell into a pit used to make charcoal. His **HOUNDS** saw him fall and stayed with him, pacing around the pit and crying and baying. Then they went looking for strangers, nipping at their clothing in order to draw them to the scene of the accident. Finally one man figured out what the hounds were up to and followed them to the charcoal pit. Unfortunately, Nicias had already burned to death, but his remains told the story of what had happened to him.

All singing animals use their tongues, as people do, but **CICADAS** produce their ceaseless chattering from their thighs. They feed on dew, and from sunrise until noon they lie silent. When the sun is hottest, though, an industrious chorus of cicadas issues a racket that descends on the shepherds and travelers and farmers who are out at that time of day. Only the male cicada loves to sing, however; the female is mute, as silent as a bashful young girl.

The Greeks say that the goddess Ergane invented weaving, but in reality nature itself taught the **SPIDER** how to spin. The spider did not imitate or study anything, and it does not have to go looking for spinning materials, but instead pulls them out of her own belly and then makes traps for insects, spreading them like nets on the sea. Not even the most skilled human weaver can compare to the spider, whose web outdoes hair in thinness.

The chroniclers praise the Babylonians and Chaldeans for their knowledge of the heavens. **ANTS** have this knowledge, too, even though they cannot look upward or count the number of days. On the first day of the month they stay indoors, close to their nests.

The male **VIPER** copulates with the female by wrapping himself around her. She allows him to do so, but when their congress is finished she treacherously repays his embraces by clamping her mouth around his head and biting it off. He dies, and she becomes pregnant, producing not eggs but live young that instantly act according to their nature: they gnaw through their mother's womb, thus avenging their father. My tragedian friends, what would your Orestes or Alcmeon have to say about this?

If you were to see a male **HYENA** this year, next year you would see a female one. The reverse is true. Hyenas share both sexes, and they marry, and having done so, they change sex year by year. This is a fact and not a fancy tale, and it makes the stories of Caeneus and Teiresias seem quaint.

The **OCTOPUS** is greedy, sneaky, and voracious, and it will eat anything. It is probably the most omnivorous creature in the sea. Here is the proof: in times of hunger, it will eat one of its own tentacles, thus making up for a lack of prey. When better times come, it grows back the missing limb. Nature thus gives it a ready meal in moments of want.

A **HORSE'S** corpse breeds wasps. As the carcass rots, they fly out of the marrow. The fastest of animals thus gives birth to the fastest of winged creatures.

The **OWL** is a wily bird. It is like a sorceress. When it is captured, it somehow manages to ensnare its hunters so that they wind up keeping it as a pet or even allow it to sit on their shoulders as if it were some sort of good-luck charm. By night it keeps them close by making incantatory noises, by which it also lures other birds. In the daytime, it keeps changing its expressions, which puzzles the birds and seizes them with terror.

Even when a skilled fisherman is after it, the **CUTTLEFISH** escapes capture by passing ink from its body and shrouding itself so that it is completely invisible. The fisherman does not see the cuttle-fish, even though it is only an arm's reach away. Homer says that Poseidon tricked Achilles by hiding Aeneas away in such a cloud.

The fish we call the **NUMBFISH** is so called because it makes whatever it touches numbly, deeply sleepy, as if drugged. The sucker-fish grabs hold of ships, which is why we call it the shipholder.

When the **HALCYON** is at rest, the sea is calm and the winds are gentle. It lays its eggs in midwinter, and a period of good weather always follows. This is why we call this time "halcyon days."

If a horse should step on a wolf's footprint, then it goes numb. If you throw a wolf vertebra among a team of horses, they will all come to an immediate halt. If a **LION** steps on ilex leaves, it goes numb. The same happens to a wolf if it steps on squill leaves, for which reason foxes throw them down into wolves' lairs. They have reason to do so, for the wolves are at war with them.

STORKS use this method of warding off bats, who make their eggs infertile: they set the leaves of the plane tree on their nests, and any bat who approaches becomes numb and can do no harm. Nature has given swallows a similar strategy, lining their nests with celery leaves to protect them from louseflies. If you throw rue on an octopus you will paralyze it. If you touch a snake with a reed, it will go numb, though if you touch it again it will regain its ability to move. The same is true for the moray eel, which goes still if touched with a sprig of fennel once but will become angry if touched again. Fishermen say that if you lay a sprig of olive on a beach, octopuses will come ashore.

ELEPHANT fat is a general remedy for many kinds of animal poisons, and if a man rubs some on his body he can withstand even the sharpest wound.

The **ELEPHANT** is frightened of rams and the squealing of pigs, and the Romans put both to use in sending the elephants of Pyrrhus of Epirus in flight, by which the Romans won a resounding victory. The elephant is also easily overcome and mollified by a woman's beauty. At Alexandria, in Egypt, it is said that an elephant competed with Aristophanes of Byzantium for the love of a garland maker. The elephant loves fragrances and is entranced by the smell of flowers and perfumes.

If a thief or a robber wants to scare off **DOGS** or make them stop barking, he grabs a firebrand from a funeral pyre and runs at them. The dogs are immediately terrified. I have heard that if a man weaves a tunic of wool from a sheep that has been attacked by a wolf, the tunic will be scratchy and annoying. "He is weaving himself an itch," the proverb puts it.

If a man wants to start an argument at a dinner party, he can do so by dropping a stone that a dog has gnawed into the wine. This will whip his guests into a fighting frenzy.

If a man sprinkles perfume on **BEETLES**, which smell nasty, then they will die. Just so, it is said that tanners, who are around foul smells all day, cannot stand the smell of perfume. The Egyptians say that vipers dread ibis feathers.

Knowledgeable fishermen catch **STINGRAYS** in the following way. The fishermen stand along the shore and dance and sing the sweetest of songs, and the stingrays are enchanted and come close to listen. The fishermen step back, little by little, luring the stingrays into their nets. The stingrays are thus caught, betrayed by song.

The **EAGLE** has the sharpest sight of all the birds. Homer tells of this fact in the story of Patroclus, when he compares Menelaus to an eagle as he searches for Antilochus so that he can send him off as a messenger to Achilles—unwelcome but needed—to announce the fate of his comrade, whom Achilles had sent out to battle, never to see his beloved homeland again. The eagle's sight serves humans in other ways. For instance, if a man whose eyesight is fading takes some eagle's gall and mixes it with Attic honey, he will see again—and with extreme sharpness of vision.

The **NIGHTINGALE** has the clearest and most tuneful voice of all the birds, and lonely places are full of its melody. They say that eating nightingale meat is a good way to stay awake. But people who eat such food are evil and stupid. Food that drives away sleep is evil, too—sleep, the lord of all gods and men, as Homer tells us.

When **CRANES** squawk, they bring on rain showers. So it is said—and also, that cranes have some sort of power which arouses women and causes them to dispense sexual favors. I take this at the word of those who have seen it happen.

The whole summer long, the **RAVEN** is parched, and it croaks—so it is said—to affirm that it is being punished. The cause is this: Apollo sent a raven out to fetch water. The raven flew out, but it came to a grain field that was not quite ripe. The raven waited until the grain was edible, ignoring its assignment. Apollo suffered from thirst, and he paid the raven back by making it suffer from the same thirst in the driest season of the year. This sounds like a fable to me, but I will repeat it here out of reverence for the god.

The raven, they say, is a sacred bird that belongs to Apollo. For this reason, it is agreed that the raven is useful in divination, and those who know about the movements of birds, the sounds they make, and the patterns of their flight are able to see into the future by listening to their croaking.

I have also heard it said that a raven's eggs will turn hair black. But anyone who uses this dye needs to fill his mouth with olive oil and keep it shut. Otherwise the dye will run down over the teeth and turn them black, and they will never be white again.

No living thing can survive the barb of a **STINGRAY**. It stabs and kills instantly, and even the most experienced fisherman holds it in the utmost fear. No one can heal such a wound. The only thing like it is the famed spear made of ash from Mount Pelion, which also inflicts untreatable wounds.

The bite of a **VIPER** and of other snakes can be counteracted with remedies, some drunk, others placed on the wound. There are spells that can undo poison, too. Only the bite of the asp, I hear, cannot be cured. For its ability to cause so much damage, the asp deserves our enmity. Even more baleful and relentless than an asp, though, is the sorceress, along the lines of Medea and Circe. An asp has to bite in order to poison, but a sorceress can kill simply by touching a person, or so it is said.

The marrow from a dead man's spine, they say, turns into a **SNAKE**. Some say that even gentle men suffer this fate, others that good and kind people die and lie at rest, their souls going on to the rewards of which wise men sing, but evildoers produce monsters in this way. I think this is a fable. Or, if we are to believe it, then it seems to me that only the corpse of a bad man would deserve this awful punishment.

The **CERASTES** is a kind of small snake that has two horns on its brow. They are something like a snail's, except they are not soft. These snakes are the enemies of all the Libyan tribes except the Psylli, who seem to be immune to their bites and are able to cure others who have been bitten. The cure is as follows: if a Psylli comes along before the poison inflames the body, then he will rinse his mouth out with water and pour water on the victim's hands, then make the victim drink the water from both. The victim recovers quickly, the infection gone. They say that if a Psylli man suspects that his wife has committed adultery or that their baby is not his, he drops the baby into a chest full of these snakes, just as a metalworker puts gold into a fire to test its purity. The snakes slither up and threaten the baby. A Psylli baby, though, will reach out and touch them, so that the snakes and the man will both know at once that he is pure of blood. The Psylli are said to be at war with other venomous beasts, particularly the malmignatte. The Libyans could be making this story up, but they should know that I am not deceived by it.

Titmice, swallows, snakes, spiders, and other creatures are at war with the **BEES**. Bees are afraid of them, and so beekeepers will try to keep these enemies at bay by using green poppies, fleabane, or other pesticides. The way to catch wasps is to hang a little cage in front of their hive in which you have put a smelt or sprat and a minnow or sardine. The wasps are greedy for food of this kind, and they will swarm out to get it. Once they are trapped, it is too late for them. Lizards have it in for bees, too, and so do land crocodiles. These can be killed by soaking grain in hellebore or the sap of mallow or spurge and then scattering it in front of the hives. These things are poison to those creatures. Beekeepers can kill tadpoles with mullein leaves or seeds. Moths are killed by lighting a candle in front of the hives and above a vat of oil. The moths fly to the light and fall into the oil and drown. Titmice become incapacitated if they eat wine-soaked grain; they fall to the ground in spasms and can easily be killed. Most beekeepers leave swallows alone out of reverence for their song; they simply keep an eye out to be sure that the swallows are not fastening their nests to the hives.

BEES dislike foul odors and sweet perfume alike, just as modest young girls do.

{ BOOK II }

When **CRANES** make ready to leave the frosts of Thrace each year, they gather at the Hebrus River. And when each has swallowed a stone for ballast against the onrushing winds, they prepare to depart for the Nile, yearning for the warmth and the food there. Just when they are about to take off in flight, the oldest crane circles around the flock three times and then falls down, taking its last breath. The other cranes bury their leader there and fly straight to Egypt, crossing the broad seas without landing or stopping to rest. They arrive when the Egyptians are just sowing their fields, and in those fields they find a feast, and uninvited they take advantage of the Egyptians' hospitality.

It is not at all strange that living creatures should be born in the mountains, in the air, and in the ocean, since matter, food, and nature make it so. But it is strange that there should be creatures that are born in fire, which we call **FIREBIRDS**, and that these should actually thrive. What is stranger still is that when these

20

creatures leave the warmth they are used to and encounter cold air, they die at once. Why they should be born in fire and die in the cold air, I leave it for someone else to explain.

There are certain creatures that we call **EPHEMERA**—that is, living for just one day—which are born in wine. When the amphora is opened, they fly out, see light, and die. Nature permits them life but then relieves them at once of life's ills, so that they do not recognize their bad luck or see the bad luck of others.

It is a fact that some men recover from the bite of an **ASP**. They do so by cutting out the poison, or surviving cauterization, or somehow preventing the spread of the venom through the radical application of medicines.

The **BASILISK** is not very large, but even the largest snake shrinks from it at the mere smell of its breath. If a man is holding a stick and the basilisk bites it, the man will die.

Of the **DOLPHIN'S** love of music and its loving nature, much has been written. The people of Corinth and Lesbos and of Ios make a theme of these characteristics. Those from Lesbos tell the story of Arion, while in Ios they tell the story of a beautiful boy who swam with the dolphins.

A Byzantine named Leonidas reports that when he was sailing past Aeolis, at a town called Poroselene, he saw firsthand a tame dolphin that lived in the harbor and treated the people as if they were his own friends. He says that an old couple adopted the dolphin and fed it the most delicious food. They had a son who had been raised alongside this dolphin, and they came to love each other as kin, which accounts for dolphins' great love of Poroselene. Moreover, the dolphin repaid the generosity of its people in this way. When the dolphin was grown up, with no need to take food from human hands, it would swim farther and farther out to sea, finding its own food and bringing back supplies for its human relatives. The boy, meanwhile, used to stand out on the rocks and call the dolphin by the name his parents had given to it, and the dolphin, no matter what it was doing—hunting, or leaping around, or racing some passing ship—would break off and go to the boy, and they would go off and play or race. Indeed, the dolphin used to let the boy win, and it acted as if it did not mind being defeated by a human swimmer.

People from elsewhere heard about the doings at Poroselene, and they came to see the dolphin, among the other things the city and island had to offer. The dolphin, then, provided for the people in still another way: as a tourist attraction.

Nature has given **DEER** an astonishing power, and that is this: deer can destroy snakes. The most powerful serpent alive cannot escape when the deer sniffs the spot where the snake has gone underground, then huffs and puffs its breath down into the hole, sucking the snake out of hiding and eating it. The deer does this most often in winter. On at least one occasion someone has ground up a deer's horn and then set the powder afire, the smoke driving all the snakes away from the vicinity. Just the smell of deer is enough to send the snakes packing.

The **STALLION** is a proud creature, for his great size, speed, height, and suppleness and the mighty crash of his hooves make him haughty and even vain. The mare is even more inclined to put on airs, mostly on account of her long mane. For one thing, she disdains being made pregnant by a donkey, thinking that she is worthy only of a great charger. Anyone who wants to breed mules, therefore, should cut the mare's mane in odd zigzags and clumps, thus making her ashamed of her appearance. Then he can put a donkey to her, far below her station though the donkey may be. Somewhere Sophocles mentions this curiosity.

I have mentioned the wisdom of **ELEPHANTS** elsewhere, and touched on how they are hunted, along with some other facts. Now I will discuss their love of music, their willingness to obey commands, and their remarkable ability to learn things that are difficult even for humans to master. Moving in a chorus, dancing, marching in time, telling different notes apart, knowing when to change pace, enjoying the song of a flute—these are all things an elephant knows how to do, and can do without making a single error. Nature has made the elephant the largest of all creatures, but this learning has made the elephant gentle and harmless. Now, if I had started to write about the elephants of India or Libya or Ethiopia, you might have accused me of making up some self-serving story, but I hunt for knowledge and love the truth, and because of that I have taken pains to say only what I have observed about them at zoos in Rome. I could say a lot more about them, too.

In any event, the elephant, when it has been tamed, is the gentlest of creatures and will obey any command. Here are some facts. Germanicus Caesar, the nephew of Tiberius, once put on a show. There were several adult male and female elephants in Rome at the time, and from them were born offspring; and when their limbs began to gain strength, a man came up to train them, teaching them in a quiet but firm voice to mind his instructions and then giving them all kinds of delicious treats, thus persuading them to give up their wildness and become tame—really, more like a human than an elephant. What they learned was not to go crazy at the sound of flutes or the beating of drums, and to be calm around the sound of marching feet and even off-key notes, and not to fear large groups of humans. Their training was human, too, in that they learned not to react in rage to a blow or to respond in anger when ordered to dance or at least to sway in time to a song. To behave with equanimity is the sign of inborn nobility in humans. Anyway, this master taught the elephants to be gentle, and they did just as he taught them, making the money spent on their education a good bargain. The elephant troupe numbered twelve, and they came into the theater on the left and right sides in two groups, mincing, swaying gently, dressed in dancers' flowing robes. They formed a line when the instructor told them to, and formed a circle, and moved offstage on command. They even sprinkled flowers on the stage floor, dancing a really wonderful rhythmic dance.

We have Damon, Spintharus, Aristoxenus, Philoxenus, and other experts on music, and we can admire them without finding their mastery of the subject to be incredible or strange. Human beings, after all, are rational creatures that are able to understand complex thoughts and logic. But that an animal which cannot speak should be able to understand meter and melody and follow the exact movements of a tragic dance without making a mistake— well, that is something truly amazing, a true blessing of nature.

What followed, however, was even more spectacular. On the theater's sand floor low couches had been set up, covered with fine pillows and fabrics that spoke to great wealth. Nearby were set gold and silver bowls full of water, and beside them lemonwood and ivory tables piled high with meat and bread and other foods. The elephants came back in, the males garbed in male clothing, the females in female raiment, and they paired off and sat down. On a signal, they ate with great delicacy and politeness, using their trunks without the slightest sign of gluttony or greediness, not like that Persian character in Xenophon's *Anabasis*. When they wanted to drink, they took in water almost daintily with their noses, spraying some on the attendants in affection, not as a challenge.

I have heard plenty of similar stories that show the astounding intelligence of these creatures. I even saw an elephant writing Roman letters on a tablet with its trunk. True, the instructor was moving the trunk, but the animal was intent at the task and kept a careful eye on the page. If you had seen it, you would have said that the animal had been taught the alphabet.

This is another strange thing about **BEARS**. A bear cannot produce a cub that it, or you, would recognize as a living being immediately after birth. It gives birth to a sort of misshapen lump, with no form or distinctive features. The mother, even so, behaves lovingly and keeps it warm, smoothing it little by little with her tongue and shaping the creature, so that after a while you can recognize the thing, finally, as a bear's cub.

The land of Ethiopia, an especially good and desirable locale that is just next door to Ocean—the place, celebrated by Homer, where the gods bathe—is the mother of the largest **SERPENTS** in existence. You have heard that some are nearly two hundred feet long and can kill elephants, and they live longer than just about any other creature. So far all the stories about them are from Ethiopia, but there are others from Phrygia about serpents that are sixty feet long. They creep out of their lairs when the sun is well in the midsummer sky, and they stretch out on the banks of the Rhyndacus River, then raise their heads higher and higher into the air and call down birds with the magical scent of their breath. The birds descend from the sky and drop straight into their mouths, enchanted. The serpents keep doing this until sundown, and then they hide and lie in ambush for flocks of sheep that are returning from pasture. They attack these flocks and slaughter them wholesale, often killing the shepherd as well to make an abundant feast.

If you hit a **LIZARD** with a stick and cut it in half, then each half will go on living, moving independently but not very well on two feet. After a while the two parts will meet again and reunite. The lizard, in other words, has one body again, although if you look closely you can see a scar that will show what has happened. Even so, the lizard will run around and do as it had been doing, apparently none the worse for the experience.

In summer, when the harvest is done and the grain is being threshed, **ANTS** assemble in large companies, marching off two abreast or in single file, sometimes even three abreast. They gather some of the grain and then march back home: some of them collect the grain and pass it off to porters, others carry it all the way, and the ones without burdens step aside for those who are laden with treasure. After a while, each one having carried its share, the ants return home and store the grains in the pits they have dug, first having cut into the center of each grain. What falls from the grain is their meal for the time being. The rest of the grain, now infertile and incapable of sprouting underground when the rains finally come, is kept safe, so that in the winter the ants will have something to eat and thus be spared from famine. Nature has provided the ants with this intelligence, as well as other gifts.

The **EAGLE** never drinks water or rests. On the contrary, it can withstand thirst and weariness, and it flies without concern through the sky, its gaze fixed on the highest heavens. At the very sound of its wings even the greatest of serpents takes cover, diving underground and thanking the stars that it has managed to escape. The male eagle tests the legitimacy of his children in this way: while they are still featherless, he arrays them facing the sun, and if one of them blinks it is cast out of the nest. The eaglets that do not blink are proven to be his, for the celestial fire of the sun is an unprejudiced and incorruptible gauge of such things.

The **OSTRICH'S** body is covered with feathers, but even so it cannot rise from the ground and fly. Still, it can attain great speeds, and when it runs it spreads its wings, which, meeting the wind, swell like a ship's sails.

The **FLY** is the most daring of creatures, but it cannot swim. When it falls into water, it drowns. But if you pick the fly's body from the water, sprinkle it with ashes, and set it in a sunny spot, the fly will come back to life.

The **STINGRAY** has a more dangerous bite than any other creature. Here is the proof: if you take a stingray's tail and touch it to a tree, that tree will immediately wither and fall over. And if the stingray touches any living creature, that unfortunate being will die at once.

The **SHREWMOUSE** has a fairly good life, and nature treats it well, except when other animals fall upon it and eat it. When it falls into a wheel rut, it cannot escape and so dies. The remedy for a shrewmouse's bite is to take some dirt from a rut and sprinkle it on the wound. The cure is immediate.

Here is another story about the Egyptian **IBIS**. The bird is sacred to the moon, and its hatching takes exactly the number of days that the moon goddess takes to wax and wane. The ibis never leaves Egypt. This is because Egypt is the wettest of all countries, and the moon is said to be the wettest of all planets. If someone tries to seize an ibis to ship it out of the country, it will fight to the death—or, failing that, starve itself to death in captivity. It walks about noiselessly, like a maiden, and never quickly. The black ibis prevents winged serpents from invading Egypt from Arabia, and the other kind of ibis catches serpents that come floating down the flooding Nile and kills them immediately. If the ibis did not do so, the Egyptians would have no defense against all the serpents that surround them.

I am told that there is a kind of **EAGLE** called the golden eagle, which some people also call the starred eagle. It is very rarely seen. Aristotle says that it hunts deer, rabbits, cranes, and farmyard geese. It is said to be the largest of the eagles, so large, in fact, that it can attack bulls. It does so in this way: as the bull feeds, its head lowered, the eagle lands on its neck and delivers a torrent of blows. The bull runs wild as if stung by a fly, galloping as fast as it can. If the bull is running on level ground, then the eagle simply follows overhead and watches it. However, when the bull approaches a cliff, the eagle spreads its wings so that the bull cannot see what lies before it, and then the bull tumbles off the cliff. The eagle then pounces on the victim and tears open its stomach, eating until gorged. It does not scavenge, though, for it prides itself on making efforts on its own account and will not allow others to help. When it can't eat any more, the eagle breathes a horrible vapor on the carcass of its prey so that no other animal can eat it. Moreover, these eagles build their nests far apart so that they do not have to argue over their prey.

EAGLES, from what I understand, love their keepers. There was, for example, an eagle that belonged to Pyrrhus, and upon the death of its master this eagle stopped eating and soon died.

Another eagle who belonged to some citizen or another threw itself on the pyre where its deceased master's corpse was burning. (Some people say that a woman, and not a man, had reared this eagle.) The eagle keeps a watchful eye on its young, and if it sees anyone or anything approaching them it will attack, beating the invader with its wings and slashing with its talons. But this punishment is comparatively mild, for the eagle does not use its beak on such occasions.

There is a kind of hawk called a **KESTREL**, and it does not drink at all. Another kind of hawk is called the orites. Both species are spellbound by the females of their kind and follow them endlessly in the manner of a lovesick man. If the female should get away, the male is quite overcome by grief and cries loudly, like someone who is heartsick.

When **HAWKS** have eye trouble they immediately find some crumbling stone wall and dig up the wild lettuce that grows along it, then hold it above their eyes while the bitter juice runs into them. This restores their vision. Doctors, I hear, also use this remedy on patients who are having eye trouble, and the remedy takes its name from the bird: "hawk medicine." These scholars have no qualms about being called followers of birds.

It is said that once a hawk at Delphi swooped down and struck the head of a man who had committed sacrilege, thus proving him guilty of that crime. Some people think that hawks are illegitimate, at least as compared with eagles of various kinds.

When spring first arrives, the hawks of Egypt choose two of their kind and send them to reconnoiter some of the desert islands that lie off the coast of Libya. When they return, they guide the rest of the hawks there. The Libyans rejoice when they see that the hawks have arrived, for they are honored guests who do no harm. Having arrived at these islands, which have been scouted and judged safe, the hawks lay their eggs and give birth to their young without having to worry about enemies. They hunt the abundant sparrows and pigeons to feed their offspring. Then, when the young ones have grown old enough to fly, the whole army of hawks goes back to Egypt and familiar surroundings.

KITES steal without end. They will even come to the market and carry off pieces of meat that have been put out for sale. However, they will leave alone meat that has been left out as a sacrifice to Zeus. Mountain kites fall on other birds and eat their eyes.

Aristotle says that **RAVENS** know the difference between a rich country and a poor one. In the former they move around in great flocks, but in the latter they travel in pairs. When their young are grown, ravens will banish them from the nest. It is for this reason that ravens gather their own food but do not look after their parents.

In Moesia, the **OXEN** have no horns. I believe that this is not because of the cold, although it is very cold there, but instead owes to some particular characteristic of the oxen themselves. Even in Scythia, after all, you will find horned oxen. I read somewhere that there are even bees in Scythia, and they do not seem to mind the cold. The Scythians even sell honey to the Moesians, and honeycombs, too, which are native to their country and not imported from elsewhere.

If I contradict Herodotus by saying this, I hope he will not be upset at me. After all, he said that he was reporting things he had researched for himself, and not just repeating stories he had heard and that cannot be checked out.

BOOK III

A **LION** will follow a Moor along his way and will drink water from the same well. I hear that lions will even come to the houses of the Moors if they cannot find prey and are starving. If the master is at home, he will drive a lion away, but if he is away and his wife is at home alone, that woman will not react violently, but instead will use words to shame the lion into restraining its hunger and its bad behavior. The lion, it seems, can understand the Moorish language, and it hears these words: "Aren't you ashamed, lion, to come here to a woman's door asking her to feed you, hoping to take advantage of her compassion and her mercy? You should be on your way to some mountainous place where you can find all the deer and antelope and other creatures that you like to eat. Now, as if you were some lazy dog, you want to be fed." And recognizing the justice of her words, the lion lowers its eyes and, abashed and ashamed, slinks away.

Horses and dogs and other creatures understand human words and cower when men threaten them. So it does not surprise me that the Moors, who live alongside these lions, are understood when they speak to them. Some Moors, after all, feed lions and share bed and roof with them as if they were their own children. So there is nothing odd in lions' understanding human speech, as described here.

This is what I have heard about the Libyan **HORSE** from accounts by the Libyans themselves. The horses are astonishingly fast and never tire, are skinny, and can put up with neglect, since their masters pay little attention to them. The Libyans do not groom their horses, comb their manes, pick their hooves clean, or braid their tails. They do not even wash them. Instead, when they finish riding, they simply dismount and turn the horses loose to graze. Of course, the Libyans themselves are skinny and dirty. The Persians, on the other hand, are proud and refined, and their horses are the same way. Horse and master alike are vain about appearances and tend toward fine garb and decorations.

It occurs to me just now to say something about **HOUNDS**. The Cretan hound is agile and can jump very high, and so it is very useful in the mountains. The Cretans themselves have the same qualities. The most aggressive of hounds is the Molossian, and the men of Molossia are fierce. In Carmania, they say, both hounds and men are wild and cannot be pacified.

If a **TORTOISE** eats a snake and then follows that up with some marjoram, it becomes immune to the snake's poison, which otherwise would be fatal to it.

I have heard it said that the **PIGEON** is the most modest of birds, sexually speaking. Pigeons never part, unless their mate dies.

PARTRIDGES, however, are wanton in their lust. They destroy their own eggs so that their females will not be too busy taking care of the nest to have sexual intercourse.

When **WOLVES** swim across a river, they do so in a way that nature has shown them to keep from being swept away. One wolf will bite the tail of the one in front of it, and so on, and then they will all swim across as a chain, so that no harm can come to them.

CROWS are very faithful, and when they mate they are monogamous, and you will never see them chasing after other crows. If one dies, its mate remains a widow. In the olden times, the people used to sing a song called "The Crow" at weddings as a way of being sure that those who have just started a family stay married. Those who pay attention to such things say that a single crow at a wedding is bad luck. The owl is the enemy of the crow by night, always looking to eat the crow's eggs. The reverse is true, for the crow, knowing that the owl cannot see by day, goes hunting for it then.

Nature has made the **HEDGEHOG** safety-minded and self-sufficient. When it goes out to gather its year's supply of food, knowing that some seasons yield poor harvests, it will climb up into a box of figs and roll around. The dry figs stick to its quills. It then will take its harvest, so to speak, back into its den and draw on this store when food is scarce.

Even the most vicious of animals will behave gently and even peacefully when it needs to. The crocodile swims along with its mouth wide open, and throughout the course of the day leeches will fall into its mouth and eventually start drawing blood, causing the crocodile pain. The trochilus acts as its doctor. When its mouth is full of leeches, it swims to a riverbank and opens its mouth, facing the sun. The **TROCHILUS** comes along and inserts its beak into the crocodile's mouth, plucking out the leeches. The crocodile stays still while the bird eats its fill. The fact that it does not eat the bird is the fee the crocodile pays for the service.

The people of Thessaly, Illyria, and Lemnos alike esteem **JACK-DAWS** as protectors and have laws providing for their feeding at the public expense. The reason is that jackdaws eat the eggs and hatchlings of the locust, great clouds of which descend to ruin crops. The number of locusts is thus slashed and the harvest saved.

In cities are found swarms of **PIGEONS**, so tame that they hop around between people's feet. In the wild they flee from people and cannot stand our company. They find courage in numbers, aware that they will not be bothered there. In the countryside it is different, with hunters and snares and the like, and there they "live fearlessly no more," to quote Euripides.

Euripides says that jealousy is a curse. Some animals, it seems, recognize this as a truth. For instance, Theophrastus says that the **GECKO**, once it has shed its skin, will turn right around and eat it up. This is selfish, for a gecko's skin is a cure for epilepsy. The deer, knowing full well that its right horn has all kinds of powers, will hide it away or bury it rather than let anyone else benefit from it. The mare knows that when she gives birth to a foal she casts love spells; this is why she bites a bit of flesh from the foal's forehead, which we call "mare frenzy." Sorcerers say that this is a powerful aphrodisiac that leads to passionate lovemaking. The mare tries to keep this secret from humans, as if not wanting to do them a favor. Is this not true?

The **SEAL**, I hear, will puke up curdled milk; otherwise it could be used as a cure for epilepsy. If this is so, then the seal is truly a nasty creature.

River-dwelling **PELICANS** are known to swallow mussels, which are then warmed up in their bellies and open up. When this happens, the pelicans puke them out and then eat the flesh. Just so, the seagull, as Eudemus observes, lifts snails high into the sky and drops them on the rocks below to open them.

Eudemus records that high up on Mount Pangaeus, in Thrace, a bear came upon a **LION'S** den while the parents were away and killed the cubs inside, which were tiny and defenseless. Just then the parents returned from hunting and saw the bear and their dead young. Grief-stricken, they chased the bear up a tree. The lioness remained under the tree so that the bear could not escape, while the lion went down the mountain and found a woodcutter. The man dropped his ax and tried to run, but the lion fawned and stretched, licked the man's face, patted his arms and head. The man relaxed, and then the lion wrapped its tail around him and guided him up the mountain, signaling that he should pick up his ax. The man did not understand, so the lion took the ax in its mouth and thrust it at the woodcutter. When the lioness saw them, she ran over and greeted the woodcutter in the same way, pointing her chin at the bear and the dead cubs. Finally, the woodcutter understood what they were trying to tell him, and so he took his ax and chopped down the tree. The bear came spilling out, and the lions tore it to pieces. Afterward the lion escorted the woodcutter down the mountain and left him just where he had been working earlier.

STORKS take care of their parents when they have grown old. There is no written law that makes them do so; they obey the law of nature. The birds love their children with the same devotion. Here is the proof: when its young are hungry, the stork will disgorge food from its own stomach and feed them. I am told that pelicans and herons do the same thing.

Storks, I am also told, migrate with the cranes to escape winter. When the cold season has passed, then the storks and cranes return to their homeland, and each of them recognizes its nest, just as a person will recognize his house after returning from a trip.

Alexander of Myndus says that when old storks are finally ready to die, they fly away to the islands of Ocean, where they are transformed into humans. This is a reward that the gods bestow for their faithful treatment of their families, since the gods want to pass on a lesson to humans about how to behave—since there is no country where people naturally behave in this way. In my view, this is not a fairy tale. If it were, why would Alexander tell it to us? He has nothing to gain by inventing a story such as this. It would not do at all for a person so intelligent to tell a lie in place of the truth, especially if no advantage or profit came from doing so.

It seems appropriate that an educated person should know this: the **CUCKOO** is a very intelligent creature, and quite skilled at solving problems. The bird knows that it cannot nest on or hatch its own eggs, because its bodily nature is cold, or so they say. When it lays its eggs, the cuckoo does not build its own nest or find one in which to brood, but instead waits until some neighbor nesting bird is off gathering food, and then puts its own eggs into that bird's nest. The cuckoo does not victimize all birds, only the lark, the ringdove, the green finch, and the pappus, which lay eggs resembling its own. If the nest is empty, then the cuckoo will leave, but if it is full of eggs the cuckoo will destroy them and substitute its own. The other birds then hatch the eggs, which are not theirs. As soon as the hatchlings can fly, they go back to their parents, for the birds that hatched them recognize that they have been tricked and behave harshly toward their young charges.

The cuckoo is seen only at the best time of year, namely, the period between spring and the rising of the Dog Star. Afterward it goes into hiding until the following year.

Lions dread **ROOSTERS**. So do basilisks. At the sound of a rooster, a basilisk will go into convulsions and then die. This is why travelers in Libya, a place that breeds many monsters, carry roosters for protection against the terrible things that can happen to them there.

In Seriphos, you will not hear a **FROG** croak, not ever. If you take a frog from Seriphos off the island, though, it will croak out an ear-splitting, most unpleasant noise. On Mount Pierus in Thessaly a lake forms each winter from meltwater, and if you throw a frog into that lake it will fall silent, even one that is quite noisy elsewhere. As to Seriphos, the people there say that Perseus came to the island after his contest with the Gorgon, and he lay down beside the lake to take a nap. The frogs, however, kept him from sleep with their loud croaking. Perseus prayed that the frogs would be silenced, and his wish was granted, condemning the frogs to eternal silence. Theophrastus, though, spoils this good story by saying that the coldness of the water is what keeps the frogs from croaking.

{ BOOK IV }

At Eryx, in Sicily, the people hold a festival that everyone on the island calls the Festival of Embarkation. They say that Aphrodite sets out during this time for Libya. As evidence, they say that the normally choking crowds of **PIGEONS** disappear at this time, and since pigeons are Aphrodite's pets—or so they say—then it stands to reason that they would go with her. After nine days, they add, a beautiful reddish pigeon comes flying over from Libya, and just so, Anacreon of Teos somewhere calls Aphrodite "rose-colored." The bird might also be called golden, in which case we have Homer, who calls the same goddess by that word. After this bird come the other pigeons in great clouds, at which time another festival, the Festival of the Return, commences.

The **WOLF** and its mate feed together, and likewise the horse and the mare. The lion and lioness do not, however, for they follow very different paths both when they hunt and when they eat.

The reason for this independence is their great strength, so that neither needs the other, as earlier writers have put it.

These animals hate one another: the tortoise and the partridge; the stork and the gull, and likewise the corncrake and the gull; the shearwater and the seagull, and likewise the **HERON** and the seagull. The crested lark hates the goldfinch, the turtledove hates the pigeon, the kite and the raven are enemies, and the hawk and the falcon agree on nothing.

Those who know about the breeding and keeping of **HORSES** observe that horses love marshes, meadows, and windy spots. Thus Homer, who in my view knew a great deal about these kinds of things, says somewhere, "Three thousand mares grazed in his pastures."

Horsemen often assert that mares are made pregnant, as they gallop, by the north or south wind. Homer knew this, too; as he says, "Boreas fell in love with them as they grazed." Aristotle, who I think was borrowing from Homer, says that mares rush in a frenzy headlong into those winds.

Eudemus tells the story of a stable boy who fell in love with a young mare, the best in the herd, as though it were a girl, and the prettiest one around. He was restrained at first, but finally he consummated his love in a very strange way. The mare already had a magnificent foal, and when the foal saw what was happening, it trampled and stamped on the stable boy and killed him. The foal even took note of where the stable boy was buried, and it went to the gravesite, dug up the body, and stomped on it and mashed it again and again.

When the new moon shows itself, **ELEPHANTS**, expressing their native intelligence in a way that we cannot explain, tear green branches from trees and then gently wave them above their heads, gazing at the goddess. It is as if they were offering Selene an olive branch in the hope that she might look kindly upon them.

The **HEDGEHOG** is a spiteful creature. When it is caught it pisses all over itself, so that the fur is unfit to use, even though it could be put to many good applications. The lynx, for its part, hides its piss away. The piss hardens into a stone that makes beautiful jewelry, and it can even be written upon.

If a **LION** eats lion's-bane, it will die. If you drop oil on an insect, it will die. Perfumes will kill vultures. If you scatter roses on beetles, you will destroy them.

Men and **DOGS**, alone of all animals, belch after they have eaten. The human heart is attached to the left breast, but in all other animals it is attached right in the center of the thorax. No bird of prey drinks or pisses, and none gathers in flocks of its own kind.

The Indians do not use hounds to hunt down rabbits or foxes. Instead, they catch a young eagle, or a raven or kite, tame it, and teach it to hunt. They attach a piece of meat to a tame rabbit or fox and then let it loose, and the bird flies after it. If the bird captures its quarry, then it is given the meat as a reward, and it makes a worthy prize. After the bird has been trained in this way for a while, it is turned loose to hunt wild hares and foxes. Such birds, Ctesias says, find prey quickly, fly down and seize it in their talons, and take it back to their trainers. The trainers keep the meat, but they give the birds the guts of whatever animals they catch.

Some people say that the **ELEPHANT** has horns, but these are properly called tusks. On each foot the elephant has five toes, though they are not separated. This is why elephants swim poorly. The elephant's back legs are shorter than the front ones. Its nose is as useful to it as hands are to a human. Its tongue is short, and its gall bladder is down near its guts, not beside its liver. I am told that the elephant gestates for two years, though some people say that it is really more like a year and a half. The elephant gives birth to a baby as big as a yearling calf that sucks at the breast with its mouth. When an elephant is crazed with the desire to copulate, it will break down walls and overturn palm trees with its forehead. (Rams use their foreheads to butt with, too.) It does not like clear water, but drinks only after it has stepped in it and stirred up mud. The elephant sleeps standing up, since lying down and getting back up takes a lot of effort. The elephant reaches the prime of life at sixty, but it lives for two hundred years. It cannot stand cold.

FOXES are clever and cunning, more so than other creatures. When a fox sees a wasps' nest, it will back into it and stick its tail in. The tail, being brushy and long, swats against the wasps, and they swarm on it and grab hold of its thick hairs. The fox will then trot off to find a big rock, then beat its tail against it, killing the wasps. It will go back to the wasps' nest as many times as it needs to, using this method, until there are no more wasps. Then the fox will stick its head in the nest and eat the honeycombs, without having to worry about being stung.

A **DOG'S** skull lacks a suture. Running, it is said, excites a dog's lust. An old dog's teeth are dull and black. Dogs have such a developed sense of smell that you could never persuade them to eat the flesh of other dogs, even if it is disguised with sauces. Dogs are susceptible to only three diseases: quinsy, rabies, and gout. Humans, on the other hand, suffer from countless illnesses. Everything a mad dog bites will die. A dog that gets gout will almost never recover its strength. A dog lives fourteen years at most, so the story about Odysseus's dog, Argos, is just a pleasantry.

Here are some of the things I have learned about **ANTS**. They are tireless, will work endlessly without shirking or excuses, and without coming up with reasons to have the day off, and they even keep working at night when the moon is full and they can see what they are doing.

Then look at people, with all their scheming and excuses for lazing about. The examples are too many to count. The Athenians have, as official holidays, the Dionysia, the Lenaea, the Festival of Urns, and Causeway Day. The Spartans have others. The Thebans have still others. Every city has such holidays, Greek and otherwise.

In India are found **INSECTS** about the same size as beetles. They are soft, with long legs, and eat the trees that produce amber. They are red. The Indians crush them and use them to dye their clothing deep red. The Persian king, it is said, loves this color, and the Persians are accordingly mad for it, it being amazingly rich compared even to the famed textiles of Sardis.

In the same part of India are found the creatures we call **DOG-HEADS**, a name that they earn because that is just how they look. They have human shape, but they are clothed in fur. They walk upright. They are no threat to humans. They howl instead of speaking, but they understand the language of the people there. They chase down wild animals very easily and cook them and eat them—cook them not with flame, but simply by tearing them apart and then letting the sun sear the meat. They keep goats and sheep for milk. I include them among the beasts, logically, since their language is unintelligible to humans.

The **LEOPARD** has five toes on each of its front paws and four on each of its back paws. The female is stronger than the male. If a leopard eats monkshood, which we call "leopard-strangler," then it will lick human excrement, which keeps it from dying.

DROMEDARIES live for fifty years, but Bactrian camels live twice as long. The Bactrians castrate the male camels used in battle, and this makes them mild and compliant without reducing their strength. If it is a female camel, they will burn the sexual organs shut.

Eudemus says that a **SEAL** fell in love with a sponge diver, and down there in its undersea garden the two would make love. This man was said to be very ugly, but the seal thought him the handsomest of all humans. This is nothing remarkable, for even human beings fall in love with ugly members of their kind, and sometimes pay no attention to the beautiful among them.

The **CYAN** bird does not like humans. It stays away from cities and houses, and it even steers clear of cultivated fields and places where there are huts or cottages. It likes lonely places, especially mountaintops and cliffs. It shuns the beautiful, settled islands and prefers places such as Skyros, barren, ugly, and unpeopled.

{ BOOK V }

It is said that the **OWL** is not found on Crete, and owls introduced to the island quickly die. Euripides, it would seem, was incorrect in writing that Polyidus saw an owl and believed that it was a sign that he would encounter the spirit of King Minos's dead son. I have determined that the Cretan histories, adding to these data, relate that Crete was honored by a gift from Zeus (he was raised there, and, remember, the island was once his hiding place): it was cleared of all dangerous creatures, neither giving birth to them nor allowing newcomers from elsewhere to survive. The island has no such creatures.

If anyone wants to try Zeus's power by introducing such creatures, well, all the animal has to do is touch Cretan soil and it will die. This poses a problem for **SNAKE** charmers, who come over from Libya with tame venomous reptiles and show them off to amazed crowds. So they bring over Libyan soil as well, laying down foreign dirt atop the Cretan earth so that the snakes are not in danger of dying. Having done this, they are able to draw crowds and entertain the foolish masses. As long as the snake remains on its native soil, it is safe. But if it goes onto Cretan dirt, then it dies, and rightly so. For Zeus's will works always, as it did with Thetis, his nurse, and it cannot fail in lesser cases than that.

The Indus River is without wild animals. The only thing that is native to it is a **WORM**, it is said, which looks like the worms that are born in and live on wood. These worms reach lengths of as much as four feet, though some are smaller. They are so big around that a ten-year-old boy cannot encircle one of them with his arms. They have a single tooth on the upper jaw, and another one on the lower jaw. Both teeth are squarish and about a foot and a half long, big enough that they can crush just about anything they bite, from rocks to animals, wild or tame. By day they live in the mud of the riverbed, and they cannot be seen. By night, however, they come onto dry land, and whatever they meet—horse, ox, or ass—they crush with their teeth and then drag into the water, where they eat the unfortunate thing, devouring every part except the stomach. On rare occasions hunger will seize them during the daytime, and then they sneak up to the surface and take a camel or ox drinking on the riverbank, dragging it down to the bottom for a feast.

These worms have skins that are as thick as two of your fingers. These are the ways in which they are hunted and caught: a fisherman will lower a strong, sturdy hook that is attached to an iron chain; to this hook is attached a long, heavy rope, and hung from the hook is a kid or lamb. A company of thirty men holds the rope, each of them armed with a spear and sword. They also keep a pile of wooden clubs close at hand in case they have to beat the worms. When a worm has swallowed the bait and gotten stuck on the hook, then the men haul it in and kill it, hanging it up in the sun for thirty days or so.

The worm exudes a thick oil that drops down into jars placed below its body, nearly a gallon's worth. The king of India is the sole recipient of this oil, and no other person is allowed to have even a drop of the stuff. The rest of the body is useless. The oil is powerful: if you wanted to burn wood and then scatter embers, you would simply pour on about half a pint of the oil and the wood would catch fire all by itself. Similarly, if you wanted to set an animal or a person on fire, you would pour on this oil and the poor thing would instantly be aflame.

So the Indian king, it is said, uses this oil to capture cities that have rebelled against him; he need not wait for battering rams or siege engines, for he can burn down those cities and capture everyone in them who is left alive. He fills ceramic jars with this oil, half a pint in each, seals them, and then uses them against the city gates. When the jars break, they set fire to the gates, and there is almost nothing that can extinguish the flames. It burns armories and armies. But it is said that you can douse the flames if you cover them with garbage.

Ctesias of Cnidus is the source of this information.

The **DOLPHIN** is said to love its own kind, and here is proof. A dolphin was captured at Aenus, in Thrace, and wounded in the process. Smelling its blood, other dolphins came racing into the harbor and jumped around, subtly threatening the fishermen. The people of Aenus, frightened, freed the captive, and the other dolphins escorted it out of the harbor.

People, on the other hand, will barely lift a finger to help a relative, man or woman, in need.

Aristotle says that the soil of Astypalaea is hostile to snakes, while that of Rhenea is hostile to martens. No **CROW** can ascend the acropolis of Athens. Say that Elis is the motherland of mules, and you will be lying.

The people of Rhegium and Locris have an agreement whereby they can enter and farm one another's lands. The **CICADAS** who live there do not share this agreement. At least, the cicadas of Locris are mute when they are in Rhegium, and the cicadas of Rhegium are silent when they are in Locris. Why this is true neither I nor anybody else, except some shiftless storyteller, can say. Only nature knows. A river runs between the two territories, and it is only a hundred feet wide at most; even so, the cicadas do not cross it. In Cephallenia, a river is the cause of both fertility and infertility among the cicadas there.

On the island of Gyarus, Aristotle writes, the **RATS** eat iron ore. Amyntas says that the rats of Teredon, in Babylonia, also eat iron ore.

I hear that on Latmus, in Caria, are **SCORPIONS** that will kill the native people but will spare travelers, injecting them with only enough venom to induce mild pain and itching. This, I believe, is a gift on the part of Zeus, the guardian of strangers.

The **FLY** is a product of nature's artistry, too, so let me not go without mentioning it here.

During the Olympics, the flies of Pisa make peace with visitors and residents alike. Even though there are countless sacrifices hanging all around, with fresh blood dripping, the flies go to the other side of the Alpheus River. They are the same in every respect as the women there, and even better behaved, for what keeps women away from the games is ritual and law, whereas the flies need no reason to leave the athletes and spectators alone. "The contests broke up, and the people dispersed," and only then do the flies come home, as if they were exiles permitted to return by some edict.

When **MICE** fall into cooking pots and cannot get out, they bite down on one another's tails. The first pulls the second up over the rim, the second pulls the third up, and so on. Nature, all wise, has taught them to come together and to help one another.

CROCODILES ambush people who draw water from the Nile in this way: they disguise themselves under driftwood and, looking up through it, swim near the shore. The people come down to the riverbank, bringing their amphoras, bottles, and jars to fill, and as they lean over to draw water from the river, the crocodiles swim up through the driftwood, overcome them, and drag them off to eat. Thus it is with the inborn evil and scheming of crocodiles.

The **PURPLE GALLINULE** is very jealous. It is also said to be absolutely devoted to its race and to love the company of other gallinules. I once heard, though, that a purple gallinule and a rooster were raised in the same household, and they ate together and walked together and even rolled around in the dust together. They became great friends. One day the head of the household sacrificed the rooster and ate the flesh. The gallinule was so distraught at the loss of its friend that it refused to eat and died of starvation.

When **GEESE** migrate over the Taurus Mountains, they are afraid of eagles, and so each of them carries a pebble in the mouth to keep the flock from making honking noises. In this way they fly over in silence and avoid attracting the eagles' attention. The goose, being of a fiery humor, loves to bathe and swim, and it prefers to eat moist food such as grass, lettuces, and other plants that generate coolness. Even if it is starving to death, though, it will not eat bay or laurel leaves and cannot be forced to do so, for it well knows that both are mortally poisonous to it.

Humans, on the other hand, know no curb on their appetites, and so they are easily taken by food and drink. There are countless examples of men who have been poisoned with drink, as was Alexander, or with food, as were the Emperor Claudius and his son Britannicus. Sometimes they drank or ate poison deliberately, but most often they were killed by some conspiracy.

Democritus says that the **LION** is the only animal born with its eyes open, and from the moment it is born it is angry in some way and ready to act on that anger. Other writers have noted that as the lion sleeps, its tail twitches, which means that it is not altogether asleep—that is, that sleep has not completely subdued it, as sleep subdues all other creatures. The Egyptians, in fact, say that the lion is always awake. I have determined that this is why they reckon that the lion belongs to the sun, for the sun is the most industrious of all the gods, always visible above or passing below without rest. The Egyptians quote Homer, who writes of the "tireless sun." The lion also exhibits great intelligence, as when it goes into a cowpen at night, scheming beforehand. Homer knew this, too; as he says, "Like cattle frightened by a lion that has come in the darkest night." The lion terrifies all the cattle with its great strength, but it seizes only one of them and devours it straightaway. When it is full, the lion saves the rest of the cows for another meal, as if it were concerned with starving. He thus breathes on the cows, and then goes away. Other lions or predators that happen by smell its breath and know the cattle belong to that lion, and they leave without attempting to rob their king of its possessions. If that lion happens to be fortunate in the hunt, it regards the cattle as old food and leaves them alone. If not, then it comes back and raids the larder from time to time.

Aristotle says that on the banks of the Hypanis River there is a sort of **FLY** that they call "one day," since it is born at dawn and dies at sunset that very day.

Nature, it would seem, has made it so that **DOGS** cure their wounds with grass. If they are bothered by worms, they eat some green grass. When they need to empty their stomachs and bowels, they eat grass, and whatever food they have not digested comes up in vomit, or some other form. The Egyptians are said to have learned the art of purging from dogs. Partridges, storks, and doves chew marjoram, then spread it on their wounds to heal their bodies. They have no need of our medicine.

Sometimes it happens that a **BEAR** will chase down a human hunter, but then abandon the hunter if they sniff at him and think he is dead. Bears, it seems, hate corpses. Mice also hate the smell of animals that have died in their holes, just as a swallow will throw another dead swallow out of the nest. Ants, guided by all-knowing nature, carry their dead out of their nests, which they keep very clean. Histories of Ethiopia, unpolluted by Greek fripperies, record that if an elephant sees another elephant lying dead, it will scoop up some dirt with its trunk and sprinkle it on the corpse, as if it were conducting a sacred ritual of some sort; failing to do so would be sacrilege. Once it has done its duty, the elephant is free to move on.

I have heard this story, too: if an elephant is dying of a wound— say, suffered in battle or inflicted by some hunter—it will gather up the grass and dirt at its feet, and then look to heaven and toss the grass and dirt in every direction with its trunk, crying and wailing in the language **ELEPHANTS** speak, as though calling on the gods to witness their suffering and the injustice of it.

Nature has given animals many different voices and languages with which to speak, just as it has done with humans. Scythians have one language, and Hindus another; Ethiopians speak one tongue, and Sacae another; the languages of Greece and Rome are not the same. And so it is with animals. Each one has a different way of speaking. One roars, another moos, another whinnies, another brays, another bleats; some get by with howling, and others barking, and others roaring. Screaming, whooping, whistling, hooting, twittering, singing—these are just some of the ways in which animals speak.

{ **BOOK VI** }

Though they are without reason, animals do not bother one another and are frequently clement. I once heard this story: A hunter had a **LEOPARD** that he had found as a cub and raised since then, caring for it as he would a friend. He loved the leopard dearly. One day he brought it a kid to eat. The leopard had already eaten and was not hungry, so it left the kid alone. On the second day, though it needed to eat, it left the kid alone again. On the third day, even though it was plainly hungry—even its voice sounded hungry—the leopard did not harm the kid, for they had been together three days, and the kid was now its friend. The leopard ate another kid, though, that was brought to it.

Yet men betray their brothers, their parents, their lifelong friends. We know of many examples.

I have already said how the **BEAR** gives birth to formless flesh and then licks it into the form of a bear. The time is now right to add this. The bear gives birth in winter, and, once she has done so, goes into hibernation. Bears cannot stand frost, and so the female waits for the coming of spring, and even then she will not bring out her cubs for three months. When she feels the onset of pregnancy, the bear behaves as if ill and goes into her lair, where she hibernates—thus, we say, hibernation is called "lair season." She goes into her lair lying down, erasing her tracks so that hunters cannot follow it. She goes in and sleeps, and she loses weight after a while. Aristotle says that the bear remains motionless for two weeks, and for the rest of her hibernation just rolls over from time to time. She spends the entire forty days of hibernation without food or drink, just licking her right paw. Because of this, her intestines dehydrate and atrophy, so that when she comes out she must immediately eat a bunch of wild arum, which makes her fart and opens up her insides so that she can accommodate food again. Then she goes and eats a bunch of ants and enjoys a fine defecation. Bears empty themselves out and fill themselves back up again, not needing doctors to tell them what to do or medicines to do it with.

In Egypt, near Lake Moeris, as we now call it, and Crocodile City, you can see a tomb to a **CROW**. The Egyptian king whose name was Mares had a tame crow that he used as a messenger. Whenever he needed to send a dispatch, the crow, which was the fastest of messengers, would fly off, having been told the destination and already having figured out where it needed to go and where it would find food and take rest along the way. Mares rewarded the crow for its good service when it died by building that tomb.

Here is more proof that animals are capable of learning. Under the Ptolemies the Egyptians taught **BABOONS** how to read, how to dance, and how to play the harp and flute. One of the baboons would go around after a performance with a bag demanding money, just as human beggars do. It has long been a matter of record that the people of Sybaris taught horses how to dance. Elephants are quick learners, as I have written. And dogs are capable of doing all kinds of things around the house once they have been trained to do so, and even a poor man can have a dog as a slave. There are people who do not even have dog slaves, such as the Troglodytes of Arabia, and the nomads of Libya, and the lake people of Ethiopia who eat nothing but fish.

After eating marjoram, a **TORTOISE** is brave enough to treat a viper with disdain. If marjoram cannot be found, it eats rue. If it fails to find either one, though, the viper will kill it.

The **HYENA**, Aristotle says, has magic in its left paw that can send someone into sleep with a mere touch. It will often steal into a stable and find an animal that is already sleeping, then put its paw on the creature's nose and keep it from breathing. The hyena then digs out the earth beneath the animal's head so that the head reclines and the throat juts out. Then the hyena will grab the animal's neck with its mouth and carry it away. It attacks dogs in this manner: when the moon is full, it will get in front of it so that the moonlight casts its shadow on the dogs. These, bewitched as if by a sorceress, are helpless. The hyena then carries the dogs off and does with them whatever it wishes.

Dogs, oxen, pigs, **GOATS**, snakes, and many other animals know when famine looms, and they can tell early on when a plague or an earthquake is impending. They can predict the weather and the year's harvest. Though they lack reason, which brings men to glory or ruin, they are quite accurate in foretelling all these things.

In the country of the people called the Judeans or Edomites at the time of Herod the Great, a massive **SERPENT** fell in love with a girl. The serpent used to come to see her and, in time, slept with her. The girl was terrified, though the serpent tried to be gentle and to approach her carefully. She escaped and kept away for a month, thinking that the serpent would forget her and find someone else. But the serpent was lonely and miserable, and it pined for the girl. It could not find her anywhere. Finally, she came back, and as if to punish her, it wrapped itself around her legs and lashed her with its tail. The serpent was probably angry that she had treated it so poorly. The god of love, who rules all the other gods, also rules animals, and this is just the sort of thing that he would do.

Aristotle tells us that **SNAKES** are aware that they have a long, narrow throat, and they are always hungry to the point of gluttony. So it is that they raise themselves up and stand on the very tips of their tails, the more easily to put food in their bodies. They have no feet, but they move very quickly all the same. One snake hurls itself at javelin speed, and this is why we call it the javelin snake.

The male **SCORPION** is terribly fierce, but the female has a gentler disposition. I have heard that there are eleven kinds of scorpions, one white, another red, another gray, and still another black; one is green, and another sort of fat, and another looks like a crab. The fire-colored scorpion is generally reckoned to be the fiercest of them all. I have read reports of winged scorpions, and other kinds that have two stingers, and one relates that somebody saw a scorpion with seven vertebrae. The scorpion does not lay eggs, but gives birth to live young. Some say that the young are produced not by mating as such but by heat, and because they live in hot places, scorpions are naturally very numerous. How they attack and sting, and what happens then, and how they kill are all things you can read elsewhere.

Nature has given scorpions a particular kind of intelligence. The Libyans, who fear scorpions and their schemes, have come up with all kinds of ways to defend themselves. They wear high boots, and their beds are away from the walls and high off the floor. The bedposts are set in buckets of water. They think all this will keep them safe, but the scorpions are clever. They will find a hole in the roof, and one will lower itself into it, letting its stinger hang down; then another will climb down and latch onto the stinger of the first, and then another will climb down and latch onto the stinger of the second, and so forth. Finally, the last scorpion scuttles down and stings the sleeping victim, then runs back up the chain, as does the next one at the bottom, and the next one, until all of them have fled the scene.

The **FOX** is very intelligent. It hatches schemes against hedgehogs of this sort. Knowing that it cannot take a hedgehog by direct assault, since the thing's quills will protect it, the fox will very gently take one in its mouth and then turn it on its back, and then rip it apart unopposed, having once rightly been very afraid of its prey.

In Pontus, the foxes hunt bustards so: They climb down into holes in the ground and then stick their tails out, so that they resemble a bird's neck. The bustards come bumbling along, thinking that they are among other birds, and the foxes rush out and attack them.

Foxes are quite skilled at catching small fish. They dangle their tails in the water, catching the little fish in their bristles. Then they go up onto dry land and shake their tails, dislodging the fish and making themselves a very fine meal.

The Thracians watch foxes to gauge whether it is safe for them to cross a frozen river. If the fox gets across without bending or breaking the ice, then the Thracians judge it safe. The fox, before going out, puts its ear to the ground and listens, and if it cannot hear flowing water it knows that the ice is solid. Then it crosses the river without concern. If the fox had heard water, it never would have gone out on the ice.

The **OCTOPUS** is a lustful fish, and it couples until all its strength has vanished, leaving it drained and incapable of swimming or hunting. When this happens, little fish and crabs, particularly the kind called the hermit crab, come down and eat the octopus. They say that this is why octopuses live for only a year and no more. The female octopus is exhausted, too, by giving birth so often.

The **BEAVER** lives both on land and in the water. By day it hides itself in rivers, but at night it wanders around on land, feeding itself with whatever it finds. Hunters pursue it avidly, and the beaver knows this, so when it feels itself in danger it will chew off its own testicles and drop them in the hunters' path, just as a wise man pursued by robbers will drop whatever treasures he is carrying in order to save his life. If it has done this once but is pursued all the same, then the beaver will stand up in the path and show the hunters that their chase is pointless, for hunters believe that a castrated beaver does not taste as good as one that is fully equipped, so to speak. Often, however, beavers will tuck away their privates and then pretend they have been castrated, tricking their pursuers and keeping their own treasures.

CATERPILLARS eat plants and quickly destroy them. They in turn are destroyed if a menstruating woman walks through the vegetable garden in which they are feeding.

The worst foes of cattle are gadflies and horseflies. Gadflies are huge, and their stinger is long and potent, and they make a sharp, buzzing sound. **HORSEFLIES** are like dogflies: their noise is louder than that of gadflies, but their stinger is smaller.

In Egypt, whenever it rains, **MICE** are born instantly. They wander across the plowed fields and eat up everything in sight, causing significant damage to crops and bringing troubles on the Egyptians, who try to catch the mice or fence off the fields or dig trenches around the crops in which they burn wood. The mice are not fooled by the traps, and they avoid them. They climb up the walls, which are smooth, and leap over the trenches. The Egyptians eventually abandon their schemes and turn to prayer, begging their gods for deliverance. The mice, it seems, dread the gods, and when this happens they retreat to some mountain or another, traveling en masse in the form of a hollow square. The youngest go first, followed by their elders, who urge them on. When the front rank stops, exhausted, then the whole phalanx stops, as does an army on the march. The people of Pontus say that the mice there do the same thing. If they sense that a house is about to collapse, then the mice will get out of it as quickly as they can run. Whenever they hear the call of a marten or the hiss of a viper, then they remove their young from their burrows and find some other hole in which to live.

The francolin hates the rooster bitterly, and the rooster returns the favor. The falcon hates the crow, and vice versa. The raven hates the **OSPREY**, and the osprey the raven. The raven and the falcon hate the turtledove, and the turtledove hates them both. I have also been told that the stork hates the bat, while the bat considers the stork an enemy. The pelican, I am told, does not like the quail, and the feeling is mutual.

In Athens, Aristotle says, a man released an old **MULE** after many years of service, but the mule loved to work and kept at it. When the Athenians were building the Parthenon, the mule did not carry any loads, but it would appear on its own and walk among the younger mules as they went back and forth, as if to encourage them, like a horse harnessed to a team. It was as if a master craftsman were taking on apprentices, someone whom old age had freed of the need to work but whose experience and knowledge benefit the young immeasurably. When the people of Athens heard of this mule's behavior, they ordered that if it came to the trough for barley or grain it was not to be denied, but instead allowed to eat its fill at the public expense, just as happened in the case of an athlete there who, in old age, was given daily meals at the Prytaneum.

Near Conopeum, as it is called, quite near the Maoetic Sea, **WOLVES** frequently keep fishermen company. If you were to see them from a distance, you would think they were just ordinary dogs. The wolves behave peacefully as long as they get a share of the day's catch. If they do not, then they will tear apart the fishermen's nets.

{ BOOK VII }

I have discovered that the **COWS** of Susa are not unfamiliar with arithmetic. This is not a far-fetched claim. Consider this story: The king of Susa has a large herd of cows, and each of them daily carries a hundred buckets of water to water his parklands. The cows do this work diligently and uncomplainingly, and not one of them shirks its duty. But if you were to ask one of the cows to carry one more bucket than the one hundred, you would not prevail, even if you tried to use force instead of gentle words. Ctesias says this.

At the foot of the great Atlas Mountains, which historians and poets alike have written about, there are great pastures and dense forests with overarching canopies, and in these places live elderly **ELEPHANTS** that have come there to spend their final years. Nature leads them there, providing them with a safe place, and the water there is pure and abundant, spilling out from springs and fountains. The elephants are not bothered, having arrived at an understanding with the barbarians that they will not be hunted; they are even considered sacred, guarded by the gods who govern woods and meadows. This story is told: A certain king wanted to kill some of the elephants because of their magnificent tusks, which become huge with the passing years. The king sent three hundred archers to do the job. They were just about at the place where the elephants lived when all of a sudden a plague struck them, and all but one of them died. He returned to the capital and told the king what had happened. This is the way that those people learned that the elephants were beloved of the gods.

In Paeonia there lives a great shaggy beast, the size of a bull, called **"ONE-EYE."** When this creature feels itself in danger from hunters, it lets forth fierce and foul-smelling excrement, and if this touches a hunter it will kill him, or so I am told.

Aristotle tells us that when **CRANES** fly inland from the sea, it is a sign that a severe storm is on the way. If those birds are flying unexcitedly, it is a sign of calm, clear weather. The same is true if they are flying without making noise. If they come in squawking, though, it means foul weather. If a shearwater calls at dusk, it means the same thing. If it flies toward the sea, rain is coming. If the weather is already stormy, then an owl's hoot means that the storm will soon clear; if the weather is fair and an owl hoots, then it means that a storm is arriving. If a raven croaks loudly and flaps its wings, a storm is coming. The same is true if a raven, jackdaw, or crow cries out in late afternoon. If a jackdaw sounds like a hawk and flies erratically, then frost and snow are coming. If white birds come in great number, heavy snowstorms impend. When ducks and shearwaters beat their wings, gales are coming. When birds come racing in from the sea, storms are coming. If a robin comes into a house or cattle shed, it is obviously seeking shelter from the coming storm. When roosters or hens strut proudly and cluck, storm weather is ahead. When birds bathe, windstorms are coming. If birds dart around during a storm, good weather is ahead. When birds gather at ponds and riverbanks, they know that a storm is on the way. When seabirds come to dry land, a heavy storm is brewing. Land birds that go to water signify good weather—if, that is, they travel silently.

The Egyptians say that the **ANTELOPE** is the first creature to sense that Sirius is rising, and it signals this by sneezing. The Libyans say that in their country, goats predict rain: when a storm is coming, they race from their pens to feed, but then turn and keep their heads pointed toward those pens until the shepherd lets them go back in.

Hipparchus, back in the day of Hiero the Tyrant, was at the theater wearing a leather jacket, and he surprised the people around him by correctly forecasting rain. Hiero, impressed, sent out a message to the Nicaeans in Bithynia, where Hipparchus came from, that they were to be praised for having him as a fellow citizen. Anaxagoras, also wearing a leather jacket, was watching the Olympic Games and predicted rain. A cloudburst soon came down, and he was praised in all Greece as having wisdom more befitting a god than a man. But, after all, an **OX**, sensing rain, lies on its right side, while in fair weather, it lies on its left side.

There are more surprises. If an ox bellows and smells the air, then a rainstorm is coming. If it eats more than its usual share, then a storm is on the way. When sheep paw at the ground with their hooves, rain is in the air. If rams mount them in the morning, the storm will arrive early. The same is true when goats huddle for protection. When pigs wander out in the grain fields, the weather will clear, and when lambs and kids frisk about, a sunny day follows. Martens and mice will make squeaking noises to signal an arriving storm, whereas if a lion is seen in the fields, it means drought. If mules bray and buck, rain is on the horizon. The same is true if they kick up dust. If large numbers of rabbits appear, the weather will be good. Most humans lack this ability to predict the weather; they know it only when it gets there.

DOGS are extraordinarily faithful. The following story says as much. During the civil war in which Emperor Galba was killed, his enemies wanted to take his head as a trophy. Countless soldiers failed at the task until they were able to kill his hound, which had been trained to protect Galba and was as loyal as a fellow soldier, fighting alongside him to the very end.

Pyrrhus of Epirus was once traveling and happened to see the body of a man who had been murdered. The man's dog stood alongside the body to make sure that no further injury would befall its master. The dog had been keeping watch for three days. When Pyrrhus learned these facts he ordered that the man's body be buried, while he fed the dog by hand, bit by bit, finally persuading it to abandon the corpse. Not long afterward the king and the dog were watching a parade of hoplites. The dog sat quietly until it saw the murderers among the soldiers, and then it leaped at them and scratched them with its claws, turning back to be sure that Pyrrhus could see what was happening. The king finally caught on, and the men were caught and tortured until they confessed.

To those who do not obey the laws of Zeus and betray their friends alive and dead, all this will sound like a fairy tale. I disagree; nature has given the animals a sense of kindness and loyalty, and even if we humans have a larger share of it, we make insufficient use of our gift. I need scarcely add examples of the terrible crimes that men have committed against their friends for money. It breaks my heart that dogs are more loyal and kinder than men.

Aristotle has told us of the industrious **MULE**, and here the story of the Athenian dog is appropriate.

A thief had been waiting through the night until the darkest, quietest hour in order to rob the temple of Asclepius. While the watchmen were dozing, he stole in and gathered up several offerings. A guard dog, however, saw him and began barking furiously. The thief ran, turning to throw stones, which had no effect. Then he threw cakes at the dog, for he had had the presence of mind to bring these things along just in case something like this happened. The dog kept on barking, following the thief all the way to his house. The dog stayed outside and barked and barked. The neighbors finally determined that the dog belonged to the temple and discovered the missing offerings, and therefore they deduced that the man in the house was the thief. He was tortured, and he confessed. The dog was rewarded, meanwhile, for being such a faithful guardian, certainly more so than the humans who were paid to watch the place.

Young **ELEPHANTS** will swim across a river, but adult ones will walk across the bottom, raising their trunks above the water, while mothers hold their newborns aloft with their tusks. The young go first to protect their elders out of piety, but they let their elders go first when eating or drinking, and need no Lycurgan laws to govern their behavior. Elephants will not abandon an elderly or infirm one of their kind, even if they are being hunted. They will fight to protect such an elephant, suffering wounds instead of escaping. As for the mothers, they would sooner die than abandon their children.

When I was a boy I knew an old woman named Laenilla. People used to point at her and tell stories, and my parents said that she had fallen in love with a slave and slept with him, and her children were called bastards as a result. They were of a noble family of the senatorial order, and the children were naturally ashamed and deeply angry with their mother for her indiscretion. Fired with lust, she accused them of hatching a plot against her. The magistrate, who was both suspicious and a coward—those being the marks of a base character—believed her. Her sons, who were blameless, were therefore put to death, while this woman kept on sleeping with her slave.

Gods of our ancestors, Artemis goddess of the childbed, goddesses of birth, daughters of Hera, when we think of the things that happen in our own time, why should we talk about Medea of Colchis or Procne of Attica?

EAGLES snatch up tortoises and then drop them on the rocks from high in the air, thus breaking their shells in order to get at their meat. In this way, I am told, Aeschylus the Eleusinian tragedian died. He was sitting on a rock, thinking and writing, probably. He was bald, and an eagle confused his pate with a rock and dropped a tortoise on Aeschylus, killing him.

WOLVES are ferocious creatures. The Egyptians say that they eat each other, even plotting murder. They gather in a circle and then start running. When a wolf becomes dizzy and starts reeling and then falls over, the others rip it to shreds and devour it. They do this whenever game is scarce. Nothing is important but feeding themselves, just as nothing but money is important to a bad man.

The **LION** knows how to exact vengeance on anyone who has done it wrong, and that vengeance may be a long time in coming; as Homer says, "He nursed his anger in his breast until he acted." Juba of Mauretania, whose son was held captive in Rome, can attest to this. He was marching to put down a rebellion in the desert, and one of the young men who was with him, noble and handsome and quite good at hunting, hit a lion wandering alongside the road with a javelin. The lion was only wounded, and the expedition was moving so quickly that the boy did not have a chance to track the lion down and kill it. A year passed, and Juba and his soldiers were returning along the same route. Though there was a whole army assembled, the lion came running up and, having nursed its rage all that time, grabbed the young man with its claws and tore him to pieces. No one even thought of killing the lion. Besides, they were afraid of the lion—and in a hurry to get home, too.

WHELKS have a king and submit gladly to its rule. This king is larger and handsomer than the others. When it comes time for the whelks to sink below the waves, the king is the first to do so; when it is time for them to surface, the king is the first to rise; when it is time to move, the king takes the lead. The king brings good luck and prosperity to any fisherman who can catch it. Indeed, if a fisherman sees some other fisherman capture a whelk king, he is gladdened, as if the luck will rub off on him. At Byzantium a prize is given to the fisherman who captures the king whelk, and everyone who is out fishing that day contributes an Attic drachma to it.

Thales of Miletus did this to a **MULE** that was misbehaving. The mule was carrying a load of salt and fell sideways while fording a river. The salt was soaked, and it quickly melted away, leaving the mule without a load. The mule was quite pleased, understanding at once the difference between hard work and loafing, and from then on it staged accidents in the river to rid itself of any load it happened to be carrying. The ford was the only route that led into the island's interior, and the freighter was losing money on every trip. Thales heard the story and told the man that it was up to him to punish the mule. He told the man to load the mule down with salt—but then pile up sponges and wool atop the sack. The mule fell over in the river as usual, but once the rest of the load was soaking wet it understood that it had made an error of judgment. From then on the mule crossed the river without incident, delivering the salt safely to its destination.

ELEPHANTS worship the rising sun, lifting their trunks toward its first rays, and this is why the sun god loves them. Ptolemy Philopater tells us this. After praying to the sun god he defeated Antiochus in battle. When the fighting was over, Ptolemy prepared an unusually splendid sacrifice, including four of the largest battle elephants in his army. The night before the sacrifice he had a dream in which the god, it appeared, was going to punish him for his novel offering. Ptolemy got up and ordered four elephant statues to be made of bronze, and these he sacrificed in place of the real elephants, hoping to win the sun god's favor again. Elephants worship the gods, while people doubt whether the gods even exist—and if they do exist, whether the gods think of us at all.

BOOK VIII

Accounts from India tell us that hunters there take thoroughbred bitches to places that are infested with wild animals, and then tie them to trees and leave them. If tigers find them and, being famished, eat them, well, that's that. However, if the bitches are in heat and the tigers have eaten, the tigers mate with them, for a full tiger thinks about nothing but sex. From this coupling, tigers, not **HOUNDS**, are born. If a tiger thus born mates with a bitch, another tiger is born, but one generation later a hound is born, since the tiger's seed weakens over time. Aristotle supports this.

Hounds that are born of tigers have no interest in chasing stags or boars, but will race at a lion as if to prove their lineage. The Indians gave Alexander, the son of Philip of Macedon, proof of the strength of such hounds. They let a stag loose, and the hound being tested sat still. They let loose a boar, and the hound did not move. They let loose a bear, but the hound was not interested.

Then they let loose a lion, and when the hound saw it, the wrath was incited in it, as Homer says, and it instantly leapt at the lion and grabbed it by the throat, strangling the lion. While this was going on, the Indian king, knowing the obstinate nature of hounds, ordered his men to cut off the hound's tail. The men did so, but the hound did not let go. Then the king ordered the men to cut off one of the hound's legs, and they did so, but the hound acted as if the leg belonged to some other creature. The king ordered them to cut off another leg, and still the hound kept at it. A third, and then a fourth, and the hound did not let go. Finally the men cut off the hound's head, but its teeth remained embedded in the lion.

Alexander was amazed, but very sad that the hound had been killed in giving testimony to its own courage. The Indian king, seeing his grief, gave Alexander four hounds of the same kind, and Alexander gave the king a fine gift in return.

Fish, it seems, can be tamed and trained, and will respond when they are called with food, as is the case with the sacred eel that lives in the fountain at Arethusa. Crassus kept a **MORAY** that he decorated with earrings and jeweled necklaces, like a beautiful girl, and when Crassus called it, the moray would swim up and take whatever food it was offered. When the moray died, it is said, Crassus grieved, and he gave it a burial service. Domitius chided him, saying, "You are a fool to mourn for a dead moray." Crassus replied, "I mourn for a dead moray, but you did not mourn for the three wives you buried."

The Egyptians say that their sacred crocodiles are tame. The crocodiles do not mind when keepers touch them and open their jaws to clean their teeth. The Egyptians add that the crocodiles have the gift of prophecy. One of the Ptolemies once called a crocodile to come take a piece of food, but the crocodile did not obey. The priests divined that the crocodile foresaw that this Ptolemy was to die soon, and thus declined to accept his gift.

DONKEYS are easily seized by wolves, and bees by bee-eaters, and cicadas by swallows, and snakes by deer. The leopard seizes prey of all kinds, but especially monkeys, with its musk.

In Ethiopia, there are **SCORPIONS** that the locals call Sibritae, and these eat lizards, asps and other snakes, cockroaches, and just about everything else that crawls on the ground. I have learned that a person who steps on their droppings will develop an ulcer.

If **WOLVES** should come across an ox that has fallen into a pond, they will encircle the bank and keep it from coming to land. Finally the ox drowns. The strongest wolf in the pack then jumps in and pulls the ox up to the bank by its tail. The second strongest grabs the strongest wolf by the tail, and the next strongest grabs the second strongest by the tail, and so on until the last wolf has joined the chain, and then they haul the ox out and feast on it. If the wolves come across a stray calf, the calf may resist when they attack, pulling against the wolf as it grabs hold of the calf's snout. The wolf then lets go, and the calf goes tumbling head over heels, whereupon the wolves jump on it, disembowel it, and devour it.

ANCHOVIES, which some people call *encrasicholi* or wolf-mouths, are tiny, abundant, and the purest white. Many other kinds of fish eat them, and when they are alarmed they rush about, clinging to their neighbors for protection. So tight is their formation that ships can run over the anchovies and not separate them, and no oar or pole can part them. If you were to stick your hand in, as if in a bag of wheat or beans, you might pull off a few strips, but not whole fish. This ball of fish is called a draft, and fishermen say that a single one can fill fifty boats.

A **SOW** recognizes the voice of a swineherd, and it answers the call even at a distance. Once some pirates beached their ship in Etruria and went inland to rustle some pigs. They stole a herd of sows and drove them aboard their ship. The swineherds saw all this but kept quiet. When the ship was well at sea, the swineherds called out loudly. The sows all moved to one side of the ship, capsizing it. The pirates drowned, but the sows made it safely to shore.

SHEEP change color depending on the river from which they are drinking. This happens during their mating season. White sheep become black, and black sheep become white. This often happens near the Antandria River and a river in Thrace whose name the inhabitants can tell you. The Scamander River turns the sheep that drink from it yellow, and for this reason the locals call it the Scamander Xanthus.

Animals have memories and are capable of showing gratitude. Once there was a good woman of Tarentum whose name was Heracleis. She was faithful and attentive to her husband. When he died, she stopped living in their home and instead spent all her time in the graveyard where he was buried. A flock of **STORKS** happened to pass overhead, and the youngest, not being strong, fell to earth and broke one of its legs. Heracleis found it and tended to the young stork, wrapping its leg in a poultice and feeding it until it was healed. The stork then flew off, but it seemed to know that it owed Heracleis a debt for its life. A year passed, and the following spring Heracleis was sitting in the graveyard. The stork, flying overhead, dropped something from its bill. It was a stone. The stone looked worthless, but she took it back to her old house. That night, she awoke to see that the stone was glowing so brightly that the entire house was illuminated. It was a brilliant gem. She searched for the stork, whose scar proved that it was the one she had healed.

It is called the **HUNTER**. Nature has given it wings. It is an ally of the thrush tribe. It is black, and it has a musical voice. It is rightly called the hunter, for it enchants the small birds that come to it with its song, casting a spell on them. It loves to listen to its own voice, but it also makes a living from its song, capturing its prey thus. If it is put in a cage, though, it will punish its captor by refusing to sing, repaying slavery with silence.

The **ELEPHANT** emerges from the womb headfirst, and it is about the size of a suckling pig at birth. A mother elephant will give birth to several, and they follow her in a chain. She will allow you to pat a newborn, knowing that no one would harm such a little creature. Even so, elephants are not easy to capture. When they are taken in a pit trap, they forget their innate wildness, though, and gladly accept food and water, and even wine.

BOOK IX

When a **LION** grows old, burdened by age, he cannot hunt. He hides himself away in caves or lairs in the jungle, and he does nothing about hunting even the weakest of his former prey, for he is self-conscious about his age and well aware of his incapacity. His young will come get him and take him out while they hunt, but leave him behind whenever they give chase to some animal. When they have successfully hunted, then they invite their old father to the feast. He comes quietly up, step by step, almost at a crawl, and meekly embraces his children, licking them, and then eats with them. No Solon had to deliver this as a law to the lions: nature, which supposedly knows nothing of law, teaches them to do these things. This is a law that is immutable.

Other birds are afraid of **EAGLES**, the king of birds, and cower when an eagle comes near. And not only that: if you take some eagle feathers and mix them into a bunch of feathers from other birds, the former will last forever, while the latter soon disintegrate.

Mice have many young, and many times. If somehow they manage to eat a little salt, then they will give birth to many more mice than is usual. When a crocodile is born, its parents watch closely. If it grabs hold of something immediately, then the parents believe it to be one of their own offspring, a worthy crocodile, and they show it all due affection. If it does nothing, but just sits in the crib, and does not snap at some passing insect or worm or lizard, then the parents will rip it to shreds as unworthy.

Here is another example of how strongly **ELEPHANTS** are attached to their young. Elephant hunters dig traps for them, capturing some, killing others. Other books discuss the size of the pits that the hunters dig, how big and deep they are, what they are lined with, and all that sort of thing. I am interested only in the elephant's affection, which is proven thus: if a mother elephant sees that her baby has fallen into a pit, she will rush over without a moment's pause and throw herself into the pit with it, crushing the baby and breaking her own neck, so that neither has to live with the pain of not having the other. Anyone who says that elephants have no feelings is a fool.

SEALS give birth on land, but in time they take the pups down to the water's edge and introduce them to the sea. They take them back to the place of their birth, then back down to the water, getting in a little bit at a time. After a while the pups become skilled at swimming, so that they can easily move about on both sea and land. Nature teaches them to love the places and things that their mothers do.

The **EAGLE** is a raptor, feeding on whatever flesh it can find. It eats rabbits, does, goslings, and many other creatures. The eagle that is called "Zeus's bird," however, eats no meat, but lives on grass instead. It has never heard of Pythagoras of Samos, but even so it is a vegetarian.

If you merely touch a **MALMIGNATTE**, it is said, then it will kill you, but painlessly. Cleopatra showed us that an asp's bite is very gentle. It is said that when the emperor's army approached, she asked the people at her banquets what kind of death would be painless. Those who had been wounded said that the sword caused pain, and drinking poison involved suffering convulsions and cramps, whereas death by an asp's bite was gentle or, as Homer says, mild. Other creatures, such as the toad, can kill just by belching.

Among humans, there are certain kinds of spells that can induce love. Male **FROGS** use a certain kind of cry to lure females, like a man singing a courting song; it is better called croaking. Frogs cannot mate in the water, and they avoid daytime encounters. But when the night falls, they come out from the water without fear and take pleasure from one another.

When frogs croak more loudly than usual, it means that rain is on the way.

Venomous animals can compound the force of their bites and stings. If a **WASP** stings a viper, for instance, then it will have a more dangerous sting ever after. If a fly does so, then its bite is bitter and painful. An asp's bite is incurable if it has eaten a frog. If a healthy dog bites a man, then the wound will be quite painful, but if the dog is mad, then the bite will be lethal. A seamstress was once mending a shirt that a mad dog had bitten and torn, and somehow she put the shirt in her mouth while she was sewing it, and she went completely mad and soon died. If a human bites another human while fasting, then the bite will be very difficult to remedy. The Scythians are said to put human blood on their arrows along with a certain kind of poison, and the blood somehow acts as a kind of primer so that the poison stays on top. Theophrastus is our witness to this.

When a **SNAKE** sheds its skin at the beginning of spring, then it also loses the mist that forms over its eyes and the sort of old age that has fallen on its vision. Rubbing fennel along the rims of its eyes, it soon rejuvenates its sight. After hibernating in close quarters underground all winter long, it cannot see far ahead of itself. But the fennel helps make its eyesight sharp again.

Along the Nile grows an herb called wolfsbane, which is very accurately named. If a **WOLF** steps on it, it will go into convulsions and quickly die. Egyptians worship the animal, and so they prevent this herb from spreading.

If a **BIRD** falls into an amphora of wine and drowns, then the wine will not be harmed; neither will anyone who drinks it. But if a bird drowns in water, then the water will putrefy and stink. If a gecko falls into wine, then, just so, it will do no harm. But if it falls into oil, then the oil will stink, and anyone who tastes it will break out in hives.

STARFISH are at war with oysters, and they feed on them. Oysters open up to cool themselves down and to feed on whatever happens to be swimming by, and when they do, a starfish will swoop down and stick one of its arms into an oyster to keep it from closing up again. This is an interesting characteristic of the starfish.

The Euphrates, which flows between Persia and Syria, is greater than all other rivers. I will discuss this elsewhere. For the moment, what matters is this: near the headwaters of the Euphrates lives a certain kind of **SNAKE** that is a deadly enemy to humans—but not all humans, only foreigners to that country. Humans who are native to the place go unharmed, but the snakes kill anyone else who comes along.

When it walks, a **LION** does not move straight ahead. Instead, it takes pains to disguise its tracks, sometimes stepping back into them and turning around, sometimes walking back and forth to erase them. This is so hunters cannot follow it and find out where its lair is and kill its cubs. These are the particular gifts nature has bestowed on lions.

I do not have the space here to list all the benefits of wormwood, which is good for lung and throat ailments—and, even better, kills intestinal **WORMS**, of which it is the sworn enemy. A worm of this kind can grow and grow and grow to monstrous proportions, and it causes all kinds of ailments that are very difficult to cure. Hippys, the historian from Rhegium, tells this story: It seems that a woman was suffering from an intestinal worm, and not even the best doctors knew what to do about it. She went to Epidaurus and prayed to the god for a cure. The god was not there at the time, but the keepers of the temple told her to lie down on the bed where cures were effected. The woman lay down, and the attendants cut off her head. One of them stuck his hand down deep inside her and pulled out a huge worm. They could not get her head back on right, though. The god arrived and, after angrily berating his acolytes for their ineptitude, fixed the problem and restored the woman to health.

As for me, Lord Asclepius, great benefactor of all humans, I would not claim that your skill was any less powerful than wormwood—that would be stupid. But in writing here about wormwood, I am reminded of your healing powers and good works. I do not doubt for a moment that this herb is also one of your gifts.

There is a certain kind of **BEETLE** that grows in the wheat fields, and on poplar and fig trees as well, as Aristotle notes. Caterpillars grow among pea plants, and spiders among bitter vetches, and the millipede, called the leek-cutter, among leeks. The cabbage caterpillar gets its name from living among cabbages. The apple tree produces a kind of caterpillar that can kill its fruit, but which helps women who are getting on in years to conceive. It is for another to tell how this works.

Sometimes it happens that if an orchard lies close to the sea, the grower will discover that in summertime **OCTOPUSES** have left the water and climbed the trees to eat the fruit. The grower will catch these octopuses and make a feast of them.

If you take a **SEA URCHIN** and tear it up, then scatter the pieces out into the sea, those pieces will come together and reform, the spiny bits all making themselves whole once again. This is by some fantastic and strange force of nature.

Here is something strange about dogs. If you approach one carrying the tail of a **MARTEN** that you have let go after relieving it of said tail, then the dog will not bark. If you hang a stone from its tail, it is said, a donkey will not bray.

In the deepest winter, when the sea is choppy and the winds howl, the **FISH** develop a fear of their home. Some of them swim down to the bottom and cover themselves up with sand to keep warm, while others take cover in the rocks. Others go down deep, for, it is said, the thrashing of the waves cannot reach down that far. At the beginning of spring, when the sky brightens and the plants begin to put out green shoots, those fish sense that the sea above them is calm, and they rise to the surface and swim close to shore, as if they had traveled from a long way away.

Aristotle, Democritus, and Theophrastus, to go from youngest to oldest, all say that **FISH** do not take their nourishment from salt water but from the fresh water that is in the ocean. Because this seems implausible, Aristotle observes that it can be proved that the sea contains fresh water by lowering a hollow wax cylinder into the water and then pulling it up after a day or two, when it will be found to be full of fresh water. Empedocles agrees that the sea contains fresh water, but not all creatures can detect it. You can read about this in his writings.

{ BOOK X }

Herodotus says that **CAMELS** have four thighbones in their hind legs, and four knees, and that their genitals point toward the tail.

SNAILS understand that birds such as partridges and herons are their enemies, and so they keep away from them. Wherever these birds gather, you will see no snails. The snails called Areiones, however, are shrewd. They come out of their shells and crawl off to feed, and then those birds come down and peck around among the shells until, discovering they are empty, they fly away. The Areiones then return to their shells and go home, having eaten well and lived to tell the tale.

The **MACKEREL** of the Black Sea behave like the Persian shah, who spends the winter at his palace at Susa and the summer at his palace at Ecbatana. These fish spend the winter in the Propontis, where it is warm, but the summer at Aegialus, where the winds are gentle.

I am told that when a chef wants to keep the stomach of a **RED MULLET** from bursting when it is cooking, he will kiss the fish on the mouth. If this is done, then the fish holds together.

The female **DOLPHIN** has breasts like a human woman, and she suckles her young with abundant milk. Dolphins swim in a body, ranked by age. The young swim in front, and after them the adults. The dolphin loves her children and protects them: first come the young, then the females, then the males, all alert and on guard, keeping an eye out on the whole school. What, O great Homer, would Nestor say, whom you call the foremost tactician among all the heroes of his time?

At a glance, the **ELEPHANT** appears to offer a huge quantity of meat, but only the trunk, the lips, and the marrow of the tusks are edible. Poisonous creatures detest elephant fat, it would seem, because if a man rubs himself down with elephant fat or burns some around his camp, they will run far away.

HAWKS are sacred to Apollo, the Egyptians say. They call him Horus in their language. Hawks, they say, are the only birds that can fly directly into the sun without suffering from its heat or glare, and they fly at incredible heights without any fear of being burned up. Some observers say that up there the hawk flies upside down. The hawk is the mortal enemy of snakes and venomous creatures. No snake or scorpion or any other kind of stinging creature can escape its gaze. It does not eat fruit or seeds, but only flesh and blood. The hawk is also lecherous. If you put its legbone up against gold, it will attract it as though the bone were made of the same material as the stone of Heraclea, which casts a spell over iron. The Egyptians say that the hawk can live for five hundred years, but I do not believe it; I am merely reporting what I have heard. Homer would seem to be saying that the hawk is indeed Apollo's favorite, for he writes,

> Down the hills of Ida he went,
> like a swift hawk, the killer of doves.

There is no such thing as a female **SCARAB**. The male spills its semen out on a pile of dung and rolls it up, keeping it warm for twenty-eight days. Then, the next day, its young come out. The Egyptian warriors of old wore rings engraved with the likeness of a scarab, the meaning being that all those who fought for Egypt were the manliest of men, since there is nothing at all feminine about the creature.

The **PIG**, a gluttonous creature, will eat its own young—and even a human corpse, if it happens upon one. This is why the Egyptians say that the pig is unclean. Right-thinking people prefer animals that have a better sense of decorum. The Egyptians thus worship storks, because storks take care of their parents in old age, and honor geese and hoopoes, which show great care for their offspring and their elders, respectively. Manetho, that supremely wise Egyptian, says that anyone who tastes sow milk will be covered with boils and leprosy. The Asians detest such illnesses, and so the Egyptians are convinced that the sow offends both the sun and the moon. They sacrifice pigs to the moon goddess at their annual festival to her, but otherwise the Egyptians have nothing to do with the animals and offer them to no other gods. The Athenians, conversely, sacrifice pigs at the time of the Mysteries, and rightly so, for pigs trample the earth and will ruin a field of grain. Eudoxus says that the Egyptians spare the sows, because they tread lightly enough to press newly sown seeds into the damp ground, keeping them safe from birds.

During winter the **RAM** will sleep on its left side, while after the vernal equinox it sleeps on its right side. At each change of the season, it changes its way of sleeping.

Some Egyptians, such as the people of Ombos, worship **CROCO-DILES** as if they were the Olympian gods themselves. If a crocodile carries a child away, the people are delirious with joy, and the mother carries herself proudly, as if she has produced a meal for a god. The people of Apollinopolis, conversely, which is a district in Tentyra, capture crocodiles in nets and hang them up in persea trees, beating and flaying them while the animals shed great tears. Then they slice the crocodiles into pieces and eat them.

The crocodile is pregnant for sixty days, and it then produces sixty eggs on which it lies for sixty days. There are sixty vertebrae in its spine, and sixty muscles that wrap around its body, and it bears sixty young and lives sixty years. I am just reporting what the people of Egypt say here. It has sixty teeth, and it hibernates for sixty days each winter, fasting. The crocodiles are accustomed to the people of Ombos, who bring them food, the heads of sacrificial animals that no human would ever eat. They throw these heads into the lake, and the crocodiles swim up and eat them. The people of Apollinopolis hate the crocodile because this is the form Typho takes. Others say that a crocodile carried away one of King Psammenitus's daughters, and in memory of that ancient tragedy the people keep up their abhorrence of crocodiles.

The Vaccaei, a people from the west, consider anyone who dies of illness or disease to be a coward, and they dishonor the corpse by burning it, whereas anyone who dies in battle is considered a hero, his body disposed of by letting **VULTURES** eat it, vultures being sacred. When Romulus studied the skies in augury and saw twelve vultures in flight, he declared that there should be twelve rods in the bundle signifying royalty, the fasces. The Egyptians say that the vulture is sacred to Hera, and they deck the heads of statues of their goddess Isis with vulture feathers and carve vulture wings over the entrances to their temples.

Beyond the last Egyptian oasis one walks for seven days across barren desert, and beyond this, on the way to Ethiopia, lies the land of the **DOGFACES**. They live by hunting gazelles and antelopes. They are black, with the heads and teeth of dogs. They do not speak, but instead produce shrill cries. Beneath their chins hang beards like those of dragons, and they have sharp nails on their hands. Their bodies are covered with hair, like dogs. They run quickly and inhabit the remotest of territories, which is why it is so hard to capture them.

A **WOLF's** neck is short, so that the wolf cannot turn its head but always has to look straight out. If it wants to look back, it has to turn all the way around. Its sight is sharp, however, and it can see things no other animal can where there is no moon. This is why the gloaming is called "wolf light," which is what Homer calls the time during which wolves hunt in darkness. The wolf, some hold, is well loved by the sun, while others say that it is Apollo who cherishes the creature. Indeed, Apollo was born after Leto changed herself into a wolf, which is why Homer writes of "the wolf-born master of the bow." There is a bronze statue of a she-wolf at Delphi to honor Leto's birth pangs. Some say, though, that the statue honors a wolf that turned up some stolen temple offerings, leading officials to the place where the thieves had buried them and then uncovering the loot with her paws.

Augeas of Eleusis once gave Eupolis the comedian a handsome Molossian **HOUND**, which Eupolis named Augeas in the giver's honor. The hound was extremely devoted to its master. Once a slave named Ephialtes stole some of Eupolis's manuscripts, and Augeas the hound tore him apart. Eupolis died young, though.

The hound howled and bayed and refused to leave the grave, and there, eventually, it died of starvation. From that day on the place was called Hound's Lament.

There are many species of **CICADA**. Those who know about these things give them names such as Ashen, after the color. I do not know why the Membrax is called that. Chirper is obvious. I have heard of the Longtail, and the Shrill Cicada, and the Barbed Cicada. These are all the names of cicadas that I can think of. If anyone has more names, please let me know.

Here are a few facts about **DOGS**. Puppies are born blind, emerging sightless from the womb. For the first three weeks they cannot see, but after the moon appears the dog has the sharpest sight of any animal alive. The Egyptians named a whole district, Cynopolis, after dogs, which they honor. They say that the name has two origins. One is that when Isis was searching for Osiris, dogs showed her the way and warded off marauding beasts. And when the Dog Star rises—this is the dog of Orion—the Nile also rises to nourish the fields of Egypt. The Egyptians honor the dog, then, because it calls up this life-giving water.

I am told that in Eryx, the site of the famous temple of Aphrodite, there is a fabulous store of gold, silver, gems, and jewelry, all gifts to the goddess. Hamilcar the Carthaginian looted this treasury, melted down the metals, and distributed coins to his army. He was punished for his crime by being tortured and crucified, while all those soldiers died horrible deaths, and Carthage, which up until that time had been prosperous, was destroyed, its people forced into slavery. These are all interesting facts, but they have nothing to do with the point of my story, which is this: The people of Eryx, and visitors to that place, sacrifice to the goddess throughout the year. The largest of the many altars there is outdoors, and all day and night a fire burns before it. At sunrise every day the fire shows no signs of aging; it forms no ash, but instead grass and dew spring up around it. Sacrificial animals come walking up to this fire without being dragged there, offering themselves to anyone who wishes to honor the goddess. The only requirement is that you pay what you are able to; a rich man will not be overcharged, and a poor man will not be turned away for having no money, but if you want to pay less than you are able to, then the animal will walk away from you, and you will not be able to offer a sacrifice.

This is one curiosity among others that I have mentioned about Eryx.

{ BOOK XI }

One day each year the people of Epirus, and the many people who come there from elsewhere, hold a festival honoring the god Apollo. The festival is both ceremonious and serious. There is a walled grove that is sacred to the god, and within this grove are **SERPENTS** that are sacred to him as well. The people of Epirus say that these serpents are descended from the python at Delphi. A priestess, a virgin, goes in alone bearing food. If the serpents look favorably on her and take the food she offers, then, the people of Epirus say, it will be a bountiful and healthful year. If the serpents scare the priestess off and refuse to take the food she brings, however, the opposite will be true—and that is just what the people of Epirus expect every year.

At Etna, in Sicily, there is a temple surrounded by a grove, and sacred **HOUNDS** keep watch over the place. They run up to people of pure hearts wagging their tails and jumping up and down, gladly. If someone approaches whose hands are stained with crime, they will bite viciously. They will merely chase away anyone who comes to the temple after a night of debauchery.

In the country of the Daunii there is a famed temple to Athena of Ilium. It is said that the **HOUNDS** there fawn on any Greeks who come to the place, but bark at foreigners.

At Curias there is a temple to Apollo, and there are also many **DEER** and plenty of hunters to chase them. The deer take sanctuary in the temple, trusting in the god to keep them safe, and the hounds there simply bay at them but do not attack.

At Icarus, an island in the Red Sea, there is a temple of Artemis frequented by wild goats, **GAZELLES**, and rabbits. If a hunter asks the goddess for permission, then he can take much game from the island, but if he does not ask he is punished in ways that you can read about elsewhere.

I have already mentioned the various kinds of **ELEPHANTS** and their characteristics. All elephants have a good memory, though. They can remember orders. For example, when Antigonus was laying siege to Megara, a female elephant was there, one of the war elephants. The keeper's wife entrusted her newborn baby to this elephant, speaking to it in the Indian language, which elephants understand. The elephant grew to love this child, and swatted away flies as it lay sleeping. The elephant refused to eat if the child were not there, and if the child were absent then this elephant, whose name was Nicaea, would threaten to act up. If the baby started to cry, then Nicaea would rock his cradle, comforting him just like a human nurse—but this was an elephant.

Animals, as I have said, can behave jealously with respect to humans. Once there was an **ELEPHANT** that caught its keeper's wife in an adulterous act. The elephant drove one tusk straight through the woman, and the other straight through her lover. Then it left the bodies in bed so that the keeper could see for himself what had happened. This happened in India, but the story traveled around the world. In the reign of Titus, it is said, the same thing happened in Rome, but there the elephant covered the lovers up with a cloak. When its keeper arrived, it pulled the cloak away with its trunk, revealing the bodies and its bloodstained tusks.

By nature, the **DOLPHIN** is in constant motion until the moment it dies. When it needs to sleep, it floats up to the surface, falls asleep, and then sinks gradually to the bottom, which wakes it up. Then it rises, falls asleep, sinks, and bumps the bottom again. It does this over and over, midway between being awake and being asleep, never once being completely at rest.

Ptolemy II, also known as Philadelphus, was once given a young **ELEPHANT**. Brought up in a Greek-speaking household, the elephant came to understand that language. Until that time, it was thought that elephants could understand only the Indian language.

A farmer was once digging out a trench to plant a vine, and he brought down his hoe on a sacred **ASP** that was crawling around underground and cut it in two. As the farmer was breaking up the soil, he saw the tail wriggling around, and then the bloody belly and head, and he went right out of his mind. All day he ran around in a frenzy, and at night he would wake up screaming that the asp was chasing him and was about to bite him. Some days he actually said that he had been bitten, groaning in agony. His illness lasted for quite a while, until his relatives took him to the temple of Serapis and begged the god to remove the asp's curse. The god took pity on the man and cured him. It is plain to see, though, that the asp had his revenge—and quickly and appropriately, too.

The Egyptians say that a live **HAWK** is beloved of the gods, and when it has died, it takes on new powers as a spirit and sends prophetic dreams. The Egyptians say that a three-legged hawk once appeared among them, and those who believe the story see no reason not to.

{ BOOK XII }

There is a bay at Myra, in Lycia, where a spring pours into the sea, and here stands a temple to Apollo. The resident priest scatters the flesh of slaughtered calves on the waters of the sea, and great shoals of **PERCH** come swimming up to feast, as if they were invited guests. The priest and the others who participate in the sacrifices take much pleasure in this, for they take the feasting of the fish to be a good omen, and they believe that the god himself has eaten by proxy. From time to time, however, the fish turn up their tails at the feast and even flick the meat back on shore, as if it were somehow bad. Then the sacrificers say that the god is displeased. The fish recognize the priest's voice, and if they respond to the sacrifice favorably those people for whom the sacrifice has been made feel that he has done well; if the fish shun the food, however, those same people are unhappy.

The Egyptians say—though I don't believe them for a minute—that in the days of the renowned Bocchoris, a **LAMB** was born with eight feet and two tails, and it could speak. They add that the lamb had two heads and four horns. Now, we can forgive Homer for granting the power of speech to Xanthos the horse; after all, Homer is a poet. Alkman the poet is also to be forgiven for following Homer's lead, for Homer sets a powerful example to all poets. But what can you say about the Egyptians who go around saying such wild things? Even though they are ridiculous, I still want to record the strange things they say about this lamb.

We all find the Egyptians ridiculous for worshipping various kinds of animals. But, in fairness, the inhabitants of Thebes, who are Greeks, worship a **MARTEN**, or so I have heard. They say that this marten was Herakles's nurse, or, if not nurse, at any rate this marten ran close to Alkmena when she was laboring to give birth to Herakles, and the sight caused her to drop the baby immediately. Herakles crawled just as soon as he was out of the womb. The inhabitants of Hamaxitos, near Troy, worship a mouse, and they say that this is why they call their god Sminthian Apollo, for the Aeolians and people of the Troad call the mouse *sminthos*, which is also the word Aeschylus uses in his play *Sisyphus* when he writes, "No, but what *sminthos* of the field is so monstrous?" In the temple of Sminthos there, tame mice are fed at public expense, and beneath the altar white mice have nests, and by the tripod of Apollo stands a statue of a mouse. I have also heard this story about the cult: Once thousands and thousands of mice came along and devoured the crops of the Trojans and Aeolians, ruining the harvest. The god at Delphi told them to sacrifice to Sminthian Apollo, and they did so, and the mice stopped eating their grain and allowed the harvest to come. Some Cretans went off to found a colony, and they asked Pythian Apollo to tell them where they should do so. The god answered through an oracle that they should found their colony where the earth-born made war against them, and they went off. When they arrived at Hamaxitos and pitched their tents, a huge army of mice swept over them and ate through their shield straps and bowstrings. The Cretans determined that these were the "earth-born" the oracle had told them about. All this talk about mice has led us into matters of theology, but no one is the worse for hearing stories like these.

DOLPHINS, I understand, are mindful of their dead and would not think of abandoning one of their number if it were to fall. When that happens, dolphins will swim underneath their dead comrade and nose him to the shore, where they are confident some man will find and bury him. Aristotle saw this himself. A company of dolphins follows behind like an honor guard, and some swim out on the flanks to protect the body lest a hungry big fish decide to swim in for a bite. Any just man, especially a music lover, will bury a dolphin, if only out of respect for their musical abilities. But those whom neither the Muses nor the Graces have touched have no concern for dolphins. So, beloved dolphins, I am afraid that you will have to pardon us savage humans, for think of this: Even the people of Athens once threw the honorable Phocion out of their city unburied, and even Olympias once lay unburied, too, even though, as she was always bragging, she was the mother of a son of Zeus's, which even Zeus admitted. The Egyptians killed the Roman Pompey, who was called "the Great" and who had in fact accomplished much, having celebrated three triumphs and having saved his murderer's father's life and restored him to the throne of Egypt. Well, anyway, they cast his headless corpse out down by the sea, the very place where humans leave dolphins to rot. And then there are men who will pickle and eat you, for such a person could not care less that his acts are hateful to the Muses, Zeus's daughters.

The Egyptians worship **LIONS**, and they named a city after them. It is worth remarking on the lions that live there. They have temples devoted to them, and lots of room to roam. People supply them with ox flesh every day, and big hunks of meat lie all over the place, which the lions devour when they feel like it, all the while being sung to in the Egyptian language. The point of the songs is something like, "Do not cast spells on those of us who come to see you," which appears to be some sort of amulet. The lions of Egypt are in fact deified, and there are whole suites of rooms that are consecrated to their use. Some of the windows of those chambers open out to the east, and others to the west, making a nice breeze. They have exercise yards, and even a wrestling arena, except they get to wrestle with a well-fed calf. The lion practices wrestling with the calf for a while, and then it slowly brings the calf down, for these lions are lazy and not used to hunting for their food, and then the lion eats until it is stuffed and wanders off to sleep.

If we take into account, as I suppose we should, the wisdom of the Egyptians, then we would assign the front parts of the lion to fire and the rear parts to water. The Egyptians do this sort of thing with their accounts and renderings of the Sphinx, after all, making her especially impressive and showing her twofold nature by joining the body of a lion to that of a young woman. Euripides writes that the Sphinx, "drawing her tail between her leonine feet, sat down," and we have the story about the lion of Nemea falling down from the moon. That is what Epimenides says, anyway: "For I am descended from the long-haired moon, who cast the lion down to Nemea, because Queen Hera told her to do so." I think we can relegate all this to the category of myth. We have talked quite enough, here and elsewhere, about the peculiarities of lions.

The **WAX MOTH** delights in fire and flies right into lamps, attracted by their brilliance. Aeschylus, the tragedian, talks about this: "I am terrified by the foolish fate of the wax moth."

When a **MOUSE** dies a natural death, without violence, then its limbs dissolve just a little bit at a time, and it takes it a good long while to die. That is the origin of the expression "like a mouse's death," meaning time-consuming, which Menander plays on in his *Thaïs*. There is another expression, "more talkative than a turtledove," which comes from the fact that turtledoves never stop chattering, both from their mouths and from their backsides. Menander, again, writes about that expression in *The Necklace*. Demetrios, the author of the play *Sikelia*, also talks about the turtledoves' talking through their bottoms.

Mice are also said to be, well, quite sex-crazed. Cratinus is cited as an authority, for in his play *Runaway Slave Girls* he writes, "Look, now, I'll send a bolt of lightning from the blue sky to punish the debauchery of Xenophon the mouse." The female mouse is even more obsessed. These are the words of Epikrates's *Chorus:* "That damned pimp conned me, swearing by the Maiden, by Artemis, and by Persephone that the little slut was a heifer, a virgin, an unbroken filly—but the whole time she was just a first-rate mousehole." I think that means she was extremely lecherous. Philemon says this: "A white mouse, when someone attempts—well, I'm ashamed to talk about it, but that blasted woman lets out such a scream that you cannot help but draw a crowd."

The **DOLPHIN** is faster and can jump higher than any other fish, and any land animal as well. It can jump over a ship. Aristotle points this out, and he explains it this way: a dolphin holds its breath just as a human diver does. Now, a human diver holding his breath will let his breath out as if it were a drawstring and his body an arrow, so that the breath shoots him up to the surface.

The **CATFISH** is found in the Meander and Lykos, two rivers of Asia Minor, and in the Strymon in Europe. It looks something like a sheatfish. Of all the fishes it is the most devoted to its children. That is, the mother ceases to pay attention to the children after giving birth to them, as human women will sometimes do, but the male takes charge of the situation and stays beside the youngsters, warding off predators. According to Aristotle, the male is capable of swallowing a fishhook without ill effect.

A frog hates and fears the water snake, and it will try to scare the snake off with its loud croaking. The **CROCODILE** hates humans and other animals with malice. When it knows where a path lies by which people come down to a river to get water for themselves or to water a horse or camel, or even to board a ship, it will take water into its mouth again and again and flood the path so that it will be slippery, making it easy to catch prey. When men or animals slip off a gangplank they fall into the water, at which the crocodile leaps up and devours them. There is much else to say about crocodiles. For one thing, they do not like many species of Egyptian plovers, which have names that are unpleasant to the ear, for which reason I will not mention them. Only the trochilus, or clapperbill, as it is known, has good relations with the crocodile, which considers the bird a friend, for it picks leeches off the crocodile's back without any harm coming to itself.

Democritus tells us that **PIGS** and dogs give birth to many young, and he says that this is because they have multiple wombs and places to receive semen. It takes the animals several bouts of copulation to fill these places up. Mules, he says, do not give birth, because they have no wombs and no place to receive seed. The mule is not a product of nature, but a sneaky trick born of the ingenuity and adulterous leanings of humans. He adds that a mare probably became pregnant after being raped by a jackass, and humans made use of the offspring and so decided that this was a good thing. The asses of Libya are bigger than most, and they are put to work mounting mares that have been shorn of their manes. Anyone who knows anything about horse breeding knows that a mare who has her glorious mane would never permit such a mate.

The same writer says that **DEER** grow horns for the following reason. He says that the deer's stomach is extremely hot, and that the veins that run through its body are very fine, while its skull is very thin, almost like a membrane. The veins at the top of its head, unlike the others, are thick. The deer's food, or at least the most nutritious part, goes through the body very quickly; the fat goes to the outside, while the solids go up through the veins into the brain, from which the horns, well wetted with juice, sprout forth. The horns keep growing and growing as the air surrounding the deer solidifies the juices. The deer's body compels the horns to grow as the juices push out, in other words. Sometimes a deer, carried along by its own momentum, will get its horns tangled up in branches. The horns drop off, but then new horns come out almost immediately, pushed along by nature.

Animals will often conceive a love for a human. For example, an **EAGLE** once raised a human baby. I will tell you the whole story to give all the relevant facts. When Seuechoros was the king of Babylon, the Chaldeans prophesied that his yet unborn grandson would one day overthrow him. This made him afraid, and he locked his daughter away in a tower and kept close watch on her. Even so, since fate has more power than even the king of Babylon, the daughter got pregnant by some unknown man. The guards, fearful of what the king would do when he found out, waited for the birth and then threw the infant from the tower. An eagle saw the baby falling and flew beneath it, caught it on its back, and took it off to some garden. The keeper of the garden saw the baby and fell in love with it and took care of it thenceforth; the baby was called Gilgamos, and indeed he became king of Babylon.

I agree, for the most part, with people who think this story is a legend. However, I have heard that Achaemenes the Persian, from whom the Achaemenid aristocracy is descended, was also nursed by an eagle.

In the country of Elam there is a shrine to the goddess Anaitis, and there pilgrims will find **LIONS** that welcome them. If you call the lions while you are eating, they will step forward courteously like guests invited to a feast, and eat whatever you offer them, and then step backward modestly and even daintily.

The **PORCUPINES** of Libya pierce anyone who touches them and cause severe wounds and pain. Even a dead porcupine can deliver quite a sharp stab, it is said.

There is a kind of monkey that lives in the Red Sea. It is not a monkey, or a fish, but a kind of cartilaginous thing. This **SEA MONKEY** looks something like a land monkey, dark and with an apish face. But the rest of its body is covered with a shell, something like a tortoise's. The rest of the body is flat, like a flounder or a ray, like a bird with its wings outstretched. When it swims, it looks like a bird in flight. But it is different from the land monkey in one important respect: its skin is blotchy, and its neck and gills are reddish. It has a very large mouth, though, just like a monkey.

Even the gods find it appropriate to take account of the nature of animals. I have heard the story of Eurysthenes and Procleos, the sons of Aristodemus, the son of Aristomachus, the son of Kleodas, the son of Hyllos, the son of Heracles, who wanted to marry. They went to Delphi to ask the god whose family they should ally themselves with in order to make a prosperous and sound marriage; it did not matter whether Greek or barbarian. The god answered, "Go home to Sparta, the way you came, and marry whomever you find once you have seen the fiercest animal carrying off the gentlest one. You will do well." The sons of Aristodemus came to the district of Kleonas, near Corinth, and spotted a **WOLF** carrying away a lamb that it had stolen from a browsing flock.

They immediately remembered the oracle, and they married the daughters of a man who lived nearby, Thersander, the son of Kleonymos.

The gods know which animal is the gentlest and which the fiercest, and so should we.

India gives birth to many and various creatures. Some of them show how bountiful and fertile the country is, but others you would just as soon live without. I have mentioned some that are precious and rare, and with luck I will say more later. But for the moment I will describe how the earth expresses the pain caused it by giving birth to **SNAKES**. The snakes of India are many and diverse. They are dangerous to humans and every other animal. Still, the country also produces herbs that counteract their bite, and the people there know which they are, and they help each other very quickly if someone is bitten to stem the swift, violent spread of the poison. The country produces these things, as I say, in great abundance to have when needed. Any snake that kills a person, say the Indians (and they say this has also been seen by Libyans and the people of Thebes, the Egyptian one), is no longer allowed to descend into the earth and creep along homeward. The earth refuses it, and rejects it as if it were an exile. From then on it wanders about, a drifter, living in the open air in all seasons, and none of its fellow snakes have anything to do with it, and its own children do not recognize their parent. This is the penalty for manslaughter, and nature has applied it even to animals, as we all can remember. This is a lesson meant to teach anyone willing to learn.

Dogs are poorer sentinels than **GEESE**, as the Romans learned. The Celts were at war with Rome, and they had sent legions reeling back and were preparing to enter the city. In fact, they captured Rome, except for the Capitoline Hill, which was too steep for them to climb. All the places that were likely to be assaulted had been fortified, and Marcus Manlius, the consul, was in charge of the Capitoline's defense. (You will remember that it was he who awarded his son laurels for bravery, but then had him put to death for desertion.) When the Celts observed that the hill was safe from open attack on all sides, they decided to wait until the middle of the night and then scale the back side, which is a rock cliff that the Romans had not defended, since they were sure no one could come up it. Marcus Manlius and the whole garrison would have been taken save for some geese that happened to be there. Dogs will shut up if you throw food to them, but geese keep on cackling and squawking even while they're being given food; this is just something they do. And so they made a huge fuss when the Celts came, and they alerted the Romans that something was up. This is why dogs are sacrificed in Rome each year, made to pay for an act of treachery long ago, but on certain feast days a goose is carried through the city on a litter to honor it.

Here are some other things to say about animals. Because they have no firewood, the Scythians use the bones of the animals they sacrifice as fuel. Among the Phrygians, anyone who kills a plow ox is killed in turn. The Sagareans of the Caspian hold **CAMEL** races to honor the goddess Athena, and their camels are very fast and good racers. The Saracori ride mules, not just to carry loads or grind grain but also to use in war, riding them into battle just as the Greeks do horses. Any mule that is particularly given to braying is sacrificed to Ares. Clearchus, the wandering philosopher, says that the people of Argos are the only ones in the Peloponnesus who do not kill snakes. If a dog comes near the marketplace during the holiday they call the Arneid, though, they kill it. In Thessaly, when a man is about to marry, he brings as a wedding sacrifice a war horse tricked out with all its gear, and then, when he has made his sacrifice and poured libations, he takes the horse by the reins and gives it to his new wife. The Thessalians would have to tell you what this means. The people of Tenedos keep a pregnant cow for Dionysus the mankiller, and as soon as it has given birth they care for the cow just as if it were a woman on her childbed. But then they dress up the calf and kill it. The man who deals the deathblow with an ax gets pelted with stones, and he has to run until he reaches the sea. The people of Eretria sacrifice wounded or crippled animals to Artemis at the village of Amarynthos.

In addition to what I have already told you, I have heard it said that the **DOGS** belonging to Xanthippos, the son of Ariphron, were most loyal to their master. When the people of Athens were boarding their ships after the Persians had set their great war against Greece in motion, and the oracle said that it was better for them to abandon their homes and sail away on their triremes, then Xanthippos left his dogs behind. But they swam alongside his ship all the way across to Salamis, where they died. Aristotle and Philokhoros tell this story.

The water of the Crathis River turns things white. **SHEEP** and cattle and every other kind of creature that drink from it go from brown or red to white. In Euboea almost all oxen are born white, which is why the poets used to call the place "of the white cattle."

Every trained and capable sculptor and painter depicts the Sphinx as having wings. I have heard that on the island of Clazomenae there was once a **PIG** with wings, and it terrorized the place. Artemon records this in his annals of the island, where there is actually a place called "the place of the winged pig." It is quite famous. Anyone who wants to think of this as a myth is welcome to do so. I would have been sorry not to mention any good story related to animals.

At Delphi the people pay homage to a **WOLF**, in Samos to a sheep, and at Ambrakia to a lioness. I should tell you why, in each case. At Delphi a wolf once tracked down some sacred gold that had been stolen and buried on Parnassus. At Samos, likewise, a sheep found some stolen gold, for which reason Mandrobolos of

Samos dedicated a sheep to Hera. Polemon tells the first story, and Aristotle the second. The people of Ambrakia have honored the lioness ever since one tore the tyrant, Phaulos, to bits and brought them freedom. Miltiades buried three mares that had won Olympic victories in the cemetery at Kerameikos, and Evagoras of Sparta also honored his Olympian horses with a great funeral.

The Ganges, the great river of India, rises from wells, and at its source it is 130 feet deep and about 9 miles wide, and it flows without mixing with other waters. But as it goes along other rivers add their waters, and the river deepens to more than 325 feet, and it spreads out to a width of nearly 45 miles. It contains islands bigger than Lesbos and Kyrnos, and it shelters huge fish, which produce oil. The river sees **TURTLES** whose shells can hold more than a hectoliter. It has two kinds of crocodile. One is more or less harmless, but the other is vicious and insatiable, and on the end of its snout it has something that looks like a horn. The people use this second kind of crocodile to punish criminals, and they throw anyone caught in an act of crime into the water, letting the beasts do the work of public executioner.

Indians and Libyans tell different stories about animals, based on what they have seen. In India, if an adult **ELEPHANT** is captured, it is usually very hard to tame and so bent on regaining its freedom that it becomes quite bloodthirsty, and if you try to tie it down it will only make the elephant angrier. The Indians try to placate such an elephant with nice food and treats, but the elephant will have none of it. So the Indians do this: they take an instrument called a *skindapsus* and play music, and the elephant listens, and in time its anger subsides and softens, and it takes notice of the food and eats. When this happens the elephant's bonds are loosened, and it eats and eats and has no thought of running away, because elephants love music.

Libyan **MARES** are just as entranced by the sound of pipes. They become gentle and stop skittering about, and they follow the herdsman wherever he leads them, and stop when he does. If he plays his pipes enthusiastically, the mares will weep with pleasure. The herdsmen of Libya use pipes made out of laurel to charm the mares. Euripides writes about "shepherds' wedding songs," and this kind of music puts the mares into heat and makes them want to copulate. This, in fact, is how they breed horses in Libya.

DOLPHINS also love songs and pipe music. There is a statue of Arion of Methymna at Taenarum, with an inscription that reads:

> Sent by the immortals, this steed saved Arion,
> Son of Kykleos, from the Sicilian sea.

Arion wrote a hymn of thanks to Poseidon that testifies to dolphins' love of music, and it is a kind of gift to them in thanks for saving him. It goes:

> Highest of the gods,

Lord of the sea, Poseidon with your golden trident,
Earth-rattler in the swelling ocean, around you
Swims a ring of dolphins, dancing, leaping,
Nimbly flashing their feet,
Snub-nosed dogs with shiny necks,
Swift-coursing, music-loving,
Nursed by the divine Nereids,
The daughters of Amphitrite,
Even the ones who carried me,
Sailing on the Sicilian sea, to Tainareum,
In the land of Pelops, placing me on their
Humped backs and darting across the waters,
Nereos's domain, where no one can walk,
When treacherous men cast me from their ship
Into the mouth of the purple sea.

So, it seems to me, to all the characteristics of dolphins that I have
already related, we ought to add a love of music.

There is an Etruscan story going around that says that the wild
BOARS and stags of Etruria are caught using nets and hounds,
which is usually the way hunting is done, but also with music—
which may be the more important tool. They do it this way: they
set the nets and other snares in a circle, and a man who is good at
playing the pipes stands in the middle and tries to play as gently
as he can, without ever hitting a shrill or loud note. The still air
carries the tune off into the forests and into the hills and canyons,
just where these animals live. When the sound reaches them they

are at first terrified, but then they are filled with inexplicable delight, and they forget all about their children and homes, even though wild animals pay great attention to these things and do not easily wander away. Still, the creatures are enchanted, and eventually they blunder into the nets, overcome by song.

{ **BOOK XIII** }

An **EAGLE**, I have heard, once told Gordias that his son Midas would be king. It did so in this way: Gordias was out plowing, and the eagle flew above him and circled overhead, and would not leave him until he had unyoked himself at the end of the day. Just so, when Gelon of Syracuse was just a boy, a huge wolf leaped into his classroom and grabbed his tablet from his hands. Gelon rose from his desk and chased the wolf, not being at all afraid but wanting to get his tablet back. As soon as he got outside, the school collapsed, killing the teacher and the other boys. It was just by luck that Gelon escaped. Curiously, the wolf saved a human's life. In one case, using an animal, the gods gave hint of a future kingdom, and in the other they saved someone from danger. This is why the gods love animals so much.

OCTOPUSES, in time, grow huge, as big as whales, and are even numbered among the cetaceans. There was one octopus at Dikaiarchia, in Italy, that grew to be monstrously huge and shunned and loathed the food it could get from the sea, and so it actually got up onto dry land and grabbed whatever it could find. It swam up through a sewer that let the waste from the city flow down into the ocean, and it came up into a house where some Spanish merchants kept their goods, mostly pickled fish in big earthen jars. It wrapped its tentacles around the jars and broke them, then ate all the fish. When the merchants came back and found the broken jars, they figured out that someone had robbed them of a great quantity of fish, but they could not figure out who did it. They saw that the doors were intact and the roof was undamaged, and that little bits and pieces of fish were all over the place. So they decided to conceal their bravest servant inside the house. He waited in ambush, heavily armed. During the night the octopus stole in again for its meal, and, like a champion weightlifter, it crushed the jars and settled in for dinner. It was a full moon, and the house was very well illuminated. The servant, who could see the octopus very clearly, decided not to take on the creature by himself, for he was a bit of a coward, really. He waited until morning and then told the merchants what he had seen. At first they did not believe him, but then they remembered how thoroughly they had been robbed, and they decided to take whatever risks were necessary to crush their enemy. Other merchants even locked themselves in the house so that they could witness this strange thing for themselves. In the evening the octopus came along and started its usual feasting, but the merchants blocked his exit. Others grabbed their spears and swords and razors and cut the tentacles just as foresters trim a cork tree. They finally overcame the octopus, but it took a long time and much effort to do so. The curious thing about it was that this all happened on dry land. Octopuses are likely to get into trouble, and they're sneaky: those are the characteristics of the creature.

A tame **ELEPHANT** who is part of a herd drinks water, but a war elephant drinks wine—a special wine, made not of grapes but of rice or sugarcane. The elephants go out and gather flowers, for they love sweet smells and are trained in fragrant meadows. An elephant can be trained to pick flowers. It will fill a basket and then wander off to take a bath, for elephants take as much pleasure in baths as luxury-loving people do. Then it will return, and if its trainer does not return the flowers it will trumpet and refuse food until finally someone brings the flowers the elephant has gathered. Then it scatters the flowers all around its stall, so that it can sleep in all that nice aroma. It seems that the Indian elephants are more than thirteen feet tall and at least six feet wide. The largest are the elephants from Prasia, and the next largest are from Taxila.

Hunting **LEOPARDS** would seem to be a Moorish custom. The people build a stone enclosure that resembles a cage, and inside it they put a piece of rancid meat. At the opening, they build a door made of woven reeds and attach a long cord to it. The animal, which likes foul smells, catches the scent wherever it is—on a mountaintop, in the forest, in some canyon—and comes running to find the feast. It knocks the door over and rushes into the trap. This is the leopard's last meal, because the Moors tug on the rope, which becomes a noose and pays the leopard well for its gluttony.

The middle of the ocean abounds in huge sea monsters, and sometimes lightning strikes them. There are others that come very close to the shore; these are called "**WHEELS**," or trochus. They swim in great numbers near Mount Athos, and in the bay off Sigeion, and sometimes on the opposite shore, near the tomb of Artachaes and Acanthus, where you can see the canal that the Persian king ordered to be dug. These monsters are shy, but they are sometimes seen at a distance, with their huge spines. If they hear oars, they will dive to the bottom of the ocean. They curl up and spin around, and this is why they are called "wheels."

The **HYENA FISH** has the same name as the hyena of land. If you put its right fin under a sleeping man, you will induce nightmares and terrifying visions. If you cut off the tail of a live tuna, release the fish, and hang the tail on a pregnant mare, she will miscarry.

If you want to avoid shaving, dissolving the flesh of a **NUMBFISH** or jellyfish in vinegar produces a good depilatory. A young man who wishes to remain beardless should rub tuna blood on his chin. What do the sorcerers of Tarentum and Etruria have to say about that, the ones who have made potions to turn men into women?

{ BOOK XIV }

In the Ionian Sea, near Epidamnus in the land of the Taulantii, is a place called Athena's Isle. There is a lagoon where the fishermen feed schools of tame **MACKEREL**. They have a kind of peace accord between them. The fish live to be very old—some of the mackerel are quite ancient. There is reciprocity in the arrangement, for the mackerel will go out swimming and find others of their kind and, since they are of the same race, then swim right among the wild mackerel, as it were. They draw up in formation, then encircle the wild mackerel and herd them into the nets of the fishermen, having traded their own freedom for full stomachs. The fishermen slaughter the strangers, while the tame fish return to the lagoon for their afternoon meal, which the fishermen soon bring to them. This happens every day.

I have already mentioned the **SEA URCHIN**. Here are a few other matters. The sea urchin is good for the stomach, helping restore lost appetite and cure digestive ailments. It is also a diuretic, medically minded people say. If you rub it on an afflicted part of the body, then it will be cured. If you burn a sea urchin, including the shell, you will make a medicine for cleaning angry wounds. If you burn a hedgehog and mix the ashes with pitch, then rub it onto the scalp, runaway hairs will come back and cure baldness. Drinking this with wine is good for the kidneys and cures dropsy. A hedgehog liver, cured in the sun, is a cure for elephantiasis.

Those who know about these things say that the tusks of a female **ELEPHANT** are worth more than the tusks of a male. In Mauretania, every ten years, the elephants drop their tusks. Stags do the same thing, though they drop their antlers every year. Elephants prefer wet, level ground, and they will root around on the ground to force their tusks to fall. They push until the tusks are buried. Then they stamp and scrape with their feet until the ground is smooth. The fertile soil soon sends up a crop of grass to disguise the site.

The elephant has two hearts. One is the seat of anger, the other the seat of peace. This is what the accounts of the Moors say. The same people say that **LYNXES**, which have snub noses and have hair on the tips of their ears, have a fearsome grip. Euripides writes somewhere, "He comes bearing on his shoulders a boar, or the hideous lynx, an ill-born, ravenous beast." Why he says "ill-born" is something for the grammarians to figure out.

Here are some more facts about the **OSTRICH**. If you kill an ostrich and wash its stomach out with water, you will see that it is full of pebbles that the bird has swallowed. These pebbles are good for human digestive troubles. The sinews and fat are good for human muscles, too.

To capture an ostrich requires the use of horses to run it down, turning in ever-smaller circles until the exhausted bird can be rounded up. Another way to capture it is this. The ostrich makes a nest by making a hole in the ground with its feet. The center of the nest is hollow, but it is surrounded by material to keep the rain off and prevent water from coming in and drowning its nestlings. The ostrich lays eighty eggs, which hatch over a period of time, not all at once. Hunters will surround this nest with a row of spears, planting them in the ground. The ostrich will go out to feed, and then, returning, will so desire to see her young that she will spread her wings and run toward the nest. The ostrich dies a terrible death on these spears. The hunters then come and take away the young, along with their dead mother.

The **DONKEYS** of Mauretania are extremely fast, at least at the start. They fly like the wind, as if borne on wings. But they tire easily, and start to pant and complain about their aching feet, and then they stop and cry great tears, not because they are about to be killed so much as in sorrow because their feet have failed them.

It is easy to catch them, riding up on horseback and lassoing them as they stand there, leading them away like prisoners of war.

The king of India eats for dessert the sorts of things that Greeks like to eat. However, Indian accounts say that he loves to eat a certain kind of **WORM** that lives in date palms. He loves eating this worm fried. I believe these accounts, which add that he also eats the eggs of swans, ostriches, and geese. Ostrich and goose eggs are well and good, but that he should destroy swan eggs—the swan being a loyal servant of Apollo, and the most musical of birds—is something, my Indian friends, of which I cannot approve.

There is a fish called the **MYRUS**, the origin of whose name I do not know. It is supposed to be a kind of sea snake. If you take out one of its eyes and wear it as an amulet, then it will cure eye inflammation; meanwhile, the myrus grows a fresh eye. You have to release the myrus alive, though, otherwise its eye will do you no good.

TORTOISES, too, come from Libya. They have a cruel look. They live in the mountains, and their shells are used to make lyres.

In Libya there is a lake of boiling water that harbors a certain **FISH** that seems to live just fine in the heat, but will die at once if you throw it into cold water.

In the sea there is a creature with a small but extremely beautiful spiral shell, born in clear water on sunken reefs, a creature called the **NERITES**. There are two stories that I have heard about this creature. I will tell them to lighten this long narrative. Now, Hesiod sings about Doris, the daughter of Ocean, who bore fifty daughters to the sea god Nereus. Homer writes about them, too. Neither poet mentions the son born after all those daughters, who figures in lots of sailors' legends. His name was Nerites, and he was the most beautiful man or god alive. Aphrodite spent a great deal of time with him in the sea and loved him, and when she was enrolled at Zeus's command as one of the Olympian gods, she asked if she could bring him along. Nerites, however, refused, saying he would rather be with his family. Then Nerites grew wings—a gift, we have to imagine, from Aphrodite. He ignored this favor, too. Aphrodite became angry and transformed these wings into a shell, and she chose Eros to take his place with her on Olympus, giving Eros the wings she gave to Nerites.

The other story says that Poseidon was Nerites's lover, and Nerites loved him in return. Poseidon used to go about in the ocean on his chariot, tearing around on the waves, and great fishes, dolphins, and tritons would come up from the deep and follow him. But so fast was Poseidon's chariot that they couldn't keep up with him. Only Nerites was fast enough to keep up.

The story has it, though, that the sun was jealous of the boy's speed, and so turned him into the shell we see today. I do not know why he was so angry, and the story does not elaborate. I would guess, in the absence of evidence, that Poseidon and Helios were rivals. Helios may have wanted the boy to go speeding around among the constellations rather than under the sea with the sea monsters. I am not going to comment any more, but instead will keep a reverential silence with respect to these two stories. I hope I have not exaggerated.

{ **BOOK XV** }

In Macedonia they have a way of catching **FISH**. Between Berea and Thessalonica the Axios River flows. There are speckled fish in it, though you will have to ask the Macedonians what they are called. These fish feed on flies that are the size of a bumblebee but are wasp-colored and buzz like honeybees. The natives call this kind of fly *hippurus*. The flies light on the stream, and so cannot escape being detected by the fish below. When a fish sees a fly, it swims up silently, lest it ripple the water and scare off its quarry. Then it swallows the fly in a gulp, in the way that a wolf devours a sheep or an eagle steals a goose from a barnyard. It then dives. The fishermen in the area know these habits, but they do not use the flies for bait, because if a human touches them they lose their scent and their wings wither, so that the fish refuse to eat them. The fishermen do this instead: they wrap their hooks in red wool, and then attach to each of the lures two of the yellowish feathers that grow below a rooster's wattles. They lower these lures into the river on lines about six feet long. The fish are so excited to see the bait that they rush up, mouths wide open, and are quickly caught.

Demostratus, who knows all kinds of fish stories and is a wonderful teller of them, says that there is a beautiful fish called the **MOONFISH**. It is small, flat, and deep blue in color. It has dorsal fins that are not sharp or brittle, but instead soft. The fins open up around the fish to form a half-circle that resembles a half-moon, whence the fish's name. The fishermen of Cyprus say so, at any rate, to which Demostratus adds that the fish is at its best when the moon is full, and it will make trees grow if a moonfish is caught and hung on the branches. If the moon is waning, though, doing so would make the tree wither. If you throw a moonfish into a freshly dug well on a full moon, then the water will never give out. If you do this when the moon is waning, the well will dry up. Depending on the time of month, if you throw one into a bubbling spring, the thing will either gush forth or disappear.

When the fishermen on the Black Sea catch **TUNA**—and, for that matter, when the fishermen of Sicily do so, for this is what Sophron says in his wonderful book *The Tuna Fisher*, and there are tuna fisheries elsewhere, too—and the tuna are firmly in the net, then these fishermen say a prayer to Poseidon, Averter of Disasters. In considering why they do this, I would ask why they give the god this epithet. In any event, they pray to Poseidon, the brother of Zeus, to keep swordfish and dolphins away from the net, since these will come and try to rescue the tuna. Swordfish have been known to cut open nets to free their captives, and dolphins do this, too.

The pearl **OYSTER** of India is caught in this way. There is a city called Perimula—at one time Soras was its ruler, when Eucratides was the king of Bactria—that is the dwelling place of the people called the Fish Eaters. These men, they say, go out with nets and make a great ring along the shore. The pearl is produced within the shells of these oysters, which have kings, just as bees do. The king stands out for its brilliant color and size, and the divers make it a point to find it, for when they do the entire colony of oysters is left without a leader and acts as if paralyzed, like a flock that has lost its shepherd. Sometimes the king escapes and hurries its tribe away, but otherwise the Fish Eaters pickle it in a jar, and in time the flesh flakes away and a pearl is left behind. The best pearls are from India and the Red Sea. They are also found in the western ocean, by Britain, though these pearls are a little duller and darker. Juba says that pearls are also found at the Bosporus, but that these are inferior to the British ones, and much inferior to the ones from India and the Red Sea.

The **MARTEN**, I have heard, was once a human being. It was a sorcerer named Marten, without any sexual restraint and lecherous. Angry, the goddess Hecate transformed him into the evil animal. I beg the goddess's pardon, for I will leave fable-spinning to others. What is certain is that the marten is vicious: it will eat a human corpse, ripping out the eyes and swallowing them. If a marten's testicles are hung on a woman, either by trickery or with her knowledge, then she will become celibate and barren. If a marten's guts are ground up by people with expert knowledge in such things and then dropped into wine, the potion will ruin friendships and break families apart. As to the sorcerers and spellmakers, let us leave them to our protector Ares to punish.

There is a snake called the **BLOOD-LETTER** that lives in rocky canyons. It is short, only a foot long. Sometimes it is dark, sometimes fiery, and it has horns on its head. It crawls along, making a little rustling noise, not frightening at all. But when it bites, the wound immediately turns dark blue, the victim has terrible stomach pains, and the belly produces huge quantities of fluid. The first night, blood flows from the nose and throat, and sometimes even the ears. This blood is mixed with bile, and the urine is bloodstained, too. Any old scars on the victim's body open up. If it is a female blood-letter, then the venom goes to the gums, and blood flows from the fingernails while the teeth fall out. This is the snake that Canobus, Menelaus's helmsman, encountered in Egypt, and when she realized how strong the poison was, Helen broke open a blood-letter's spine and extracted some. What she was going to do with the venom, I cannot say.

There seems to be a kind of natural, mysterious bond between lions and **DOLPHINS**. One is the king of the land and the other the king of the sea, but it is more than that. When a lion is old and begins to fall apart, it will eat a monkey as medicine. When a dolphin is old, then it seeks the equivalent, and indeed there is something called a sea monkey that serves the purpose.

There is a region near Thessalonica, in Macedonia, called Nibas. The **ROOSTERS** there do not have the ability to crow, and they are silent. There is a local proverb that is used to mean that something is not going to happen: "You will have such-and-such when Nibas crows."

When Alexander invaded India and caused such a panic, he encountered among many other strange animals a **SERPENT** that lived in a cave. The Indians regarded it as sacred and begged Alexander to keep his troops from attacking it. Alexander agreed. His army went by the cave, and the serpent heard the noise—for, as you know, serpents have the keenest hearing and sight of all animals—and made such a hiss and roar that the soldiers were thrown into a panic of their own. The serpent is said to have been a hundred feet long, and not all of it could be seen, for it stuck only its head outside the cave. Each of its eyes was the size of a large Macedonian shield.

HORSES that drink from the Cossitinus River, in Thrace, become wild. The river empties out in Abdera in Bistones Lake. Here, as you know, Diomedes the Thracian had his palace, and he owned wild horses. This was one of the labors of Hercules. The same thing happens to horses that drink from the spring at Potniae.

This is not far from Thebes. They say that the people of Oraea and Gedrosia feed their horses fish; the Celts feed their horses and cows fish, too. In their country, the horses flee from the scent of humans and race south, especially when the south wind blows. There are those who say that the people of Macedonia and Lydia also feed their horses fish—and their sheep, too. In Moesia, when stallions cover mares, pipers play along, providing mood music, as it were. The mares become pregnant and produce beautiful foals. They say that if an old horse produces offspring, they will be weak and have bad legs. They say that stallions live thirty-five years. But Aristotle says that a horse once lived seventy-five years.

The Pygmies have their own form of government, but owing to some weakness in the male line, a woman became queen and ruled over them. Her name was Gerana, and the Pygmies worshipped her as if she were divine. She became quite full of herself as a result, and she did not pay the real goddesses proper respect. She even said that Hera, Athena, Artemis, and Aphrodite were nowhere near as beautiful as she. This is the kind of thing that will yield disaster, and Hera changed Gerana into an ugly bird called the **CRANE**, which wages war on the Pygmies even today simply because they spoiled Gerana and drove her to madness.

{ **BOOK XVI** }

In India, as I have mentioned, there are **PARROTS**. I have some things to add about them here. There are three kinds, I hear, and all of them learn to speak, and their power of speech grows just like that of human children. In the forest, they make bird noises, and not articulate sounds; there they are ignorant and cannot speak. India also has the largest peacocks in the world, along with green-colored pigeons that, if you did not know better, you would say were parrots. They have beaks and legs that in color are the same as those of Greek partridges. The roosters there are huge, and their combs are not red, as they are in our country, but of many colors, like bouquets of flowers. Their tail feathers do not curl up in a circle but are instead straight, trailing behind them like a peacock's plumage. Their wings are golden, with hints of emerald here and there.

In India there is another bird, about the size of a starling, that is also of many colors and quite talkative, more so and more intelligent than the parrot. It cannot stand to be imprisoned, and it would rather starve than live in captivity. The Macedonians who settled in India in the towns that Alexander founded in Bucephala and that region, and in Cyropolis and elsewhere, call this bird **CERCION**, which owes to the bird's shaking its tail—its *cercos*—just like a wagtail does.

The Indian **HOOPOE** is twice as large as the hoopoe of our country, and much prettier. Just as Homer says that the trappings of a horse are a king's glory, so the hoopoe is the great glory of the king of India. He always carries one, and, enchanted by the bird's beauty, he spends hours staring at it.

The Brahmins tell a story about this bird. It seems that a son was born to another Indian king, and this boy's brothers grew up to be outlaws inclined to great violence. They showed the king and queen no respect, and they scorned their young brother. The parents, in despair, went off into exile, taking the boy with them. It was a hard journey, and the parents died en route. The son split open his own head with a sword and buried his parents inside his brain. The Brahmins say that Helios was so overcome by the boy's act that he transformed the prince into a bird of surpassing beauty that lived for a long time. From the top of his head springs a crown, as if to proclaim the events of the boy's journey.

The Athenians tell a similar story about the lark, which Aristophanes incorporated into his play *The Birds*, writing:

> No, you were ignorant and lazy and did not read your Aesop,
> who tells us that the lark was the first bird to be born, before
> the earth itself, and that its father soon fell sick and died.
> There was no earth, and so the father's corpse lay stretched
> out for five days, until the lark, not knowing what else to do,
> buried it inside its own head.

So, it would appear, the story from India about a different bird spread around the world and reached the Greeks. The Brahmins say that it was a very long time ago that the hoopoe, while still a human being, did this for his parents.

The **SAND PARTRIDGE** is found near Antioch in Pisidia. It eats stones. It is black, with a red beak, and smaller than the regular partridge. It cannot be tamed, unlike the other partridge, but remains wild. It is not big, but it is much nicer to eat than the other partridge, with firmer flesh.

Among the Prasii, who live in India, is a race of **MONKEYS** possessed of human intelligence. This monkey is as big as a Hyrcanian hound, and it has a forelock that looks fake but is real. It has a satyrlike beard, and its tail grows long, like a lion's. Its body is white, except the head and the very tip of the tail, which are red. It is mild-mannered and tame. It lives in the forest and eats wild food, but it comes in great numbers to the outskirts of Latage, an

Indian city, and feeds on boiled rice that the king prepares for its kind every day. When this monkey has eaten its fill, it goes back to the forest, taking care not to damage any human property that it encounters.

In India there is a **GRASS-EATING ANIMAL** twice the size of a horse. It has a thick black tail, with hairs finer than those on a man's arm. Indian women prize it and braid this animal's hair into their own. Each hair can grow to a length of six feet, and perhaps thirty will come from a single root. This creature is very shy, and if it senses that someone is looking at it, it will run away as fast as it can, yearning to escape. Horsemen with very fast hounds are the only men who can hunt it. If this creature realizes that it is about to be caught, it will hide its tail, then turn around and face its attacker, reasoning that without a tail it is no longer of interest. This trick does not work, however, for the hunter will kill it and cut off its tail, the object of the hunt. The hide is then stripped, but the carcass is left behind, because the Indians do not eat the flesh of this animal.

We have intelligent animals in our country, of course, but fewer than they have in India. In that country, for example, live the elephant, the parrot, the sphinx ape, and satyrs. The Indian **ANT** is also very clever. Our ants construct burrows and lairs underground, digging out the earth and working hard at their mines, so to speak. The Indian ants, though, build little houses of the material that they assemble, and these are not down in low-lying areas, but higher up where the ground is not so easily flooded.

They build the equivalent of Egyptian catacombs or Cretan labyrinths, a great maze that is hard to follow, and all that can be seen from the ground is a single hole through which they come and go, carrying seeds into their storehouses. The rivers there flood all the time, and as a result of their construction the ants are able to survey the land all around them when the floods do come. The dew binds the salty soil, and they are dressed, so to speak, in a fine tunic of frost from that dew, so that the mound is shielded from the water, and the river mud leaves behind a dense thatch of weeds for further protection. Juba wrote about the ants of India, too. For now, this is all that I have to say about them.

In the country of the Ariani, in India, is a plutonian chasm. At the bottom, too far down to see, there is a great maze of catacombs, hidden corridors, and footpaths, which go deep into the earth. No one among the Indians knows who built this, or when; I have not done much research myself. The Indians bring here some thirty thousand **SHEEP**, goats, cows, and horses, and any of them who has been troubled by a dream, vision, or augury sacrifices whatever animal he can afford. The victims, though, come there without being driven, as if by their own will, drawn by some mysterious power. They leap into the chasm unbidden, into the silence, far from the mooing, baaing, bleating, and neighing up above, though they make those noises, too. Anyone who walks along the rim can hear those noises from a long way away. The sounds never stop, for the Indians make sacrifices there every day. No one knows whether it is only the sounds of the most recent victims that can be heard, or whether all the ones who have jumped down there continue to call for a long time afterward. This is a very remarkable thing about that country.

In India, deep in the interior of the country, stands an unscalable barrier of mountains that contain many kinds of wildlife—just as many as we have in our country, but truly wild. It is said that the sheep, the goats, the cows, and even the dogs there are wild, and they come and go as they please, without submitting to any human master. The histories of India say that these animals are innumerable, and the Brahmins say the same thing. One of these animals is called the **CARTAZONUS**, which has one horn and has light hair and a mane like a horse's. Its feet have no toes, like an elephant's, and it has a tail like that of a pig. Between its eyebrows, a horn grows. The horn is curly, black, and very sharp, or so it is said. It has an angry, frightening cry. When other animals approach, it responds gently, but it is quick to fight its own kind, whether male or female—and the fight is to the death. It is very strong, but especially in the horn. It roams the meadows alone, though at mating time it becomes gentler and sometimes grazes alongside a female. But after it has impregnated the female, it becomes wild again and goes off to live alone.

The offspring, it is said, are captured and taken off to live at the palace of the king of the Prasii, and they fight each other in exhibitions there. No one can recall a full-grown cartazonus's ever having been captured, though.

The **HORSE** is a docile creature. Consider this: I have heard it said that the people of Sybaris, in Italy, spent a great deal of energy on the fine art of luxurious living. They knew nothing about anything other than idleness, indolence, and extravagance. To tell all about how they lived would be a very long story, but the following tale says it all. The Sybarites had trained their horses to dance in time to pipe music at dinnertime, as entertainments for the banquet. When the people of Croton went to war with the Sybarites, knowing this, they did not use trumpets but instead assembled pipers and snuck up to the Sybarite lines. When the enemy was within range of their archers, the Crotonian pipers started playing dance music. The Sybarite horses shook off their riders and started to dance up and down as if they were at some bacchanalia, throwing the Sybarites into confusion and losing the war for them.

In his history of India, Ctesias says that a people called the Dogmilkers keep a kind of **HOUND** similar to the Hyrcanian breed. Ctesias says that after the summer solstice herds of wild cattle come by the Dogmilkers' country, wild and savage, like a swarm of bees that has been riled up, and they do all kinds of destruction. The Dogmilkers, otherwise defenseless, sic these hounds on the cattle, and the hounds make short work of them, leaving mountains of meat out on the plain. The Dogmilkers eat as much as they want to of the choicest cuts, leaving the rest for the hounds. The arrangement works out well for man and hound. In other seasons the Dogmilkers use the hounds to hunt other animals, and besides that, they milk the bitches, for which reason they are called what they are—for they drink hounds' milk just as we drink the milk of ewes and goats.

I have already mentioned that elephants are terribly afraid of **PIGS**. Antigonus once besieged the city of Megara. The Macedonians coated some pigs with pitch, set them afire, and turned them loose, and the pigs, shrieking in pain and panic, went tumbling into the elephant cavalry and set the elephants in panic in turn. The elephants, though highly trained, would not obey orders afterward. It may be that elephants simply cannot stand pigs in general, or they are afraid of their screaming and squealing. Whatever the case, elephant trainers now know to keep pigs alongside young elephants to inure them and teach them not to be afraid.

In Metropolis, near Ephesus, there is a lake. Alongside it at one point is a great cave, and inside this cave lives an incredible number of **SNAKES**, huge and with terrible fangs. It is said that the snakes can leave the cave and swim in the lake, but that if they try to go farther huge land crabs fall on them and kill them, seizing them with their claws. The snakes stay put as a result, for they are afraid of the crabs. It would have been impossible for people to live in the vicinity had the crabs not concerned themselves with attacking the snakes—for otherwise they would have attacked the humans.

In his book *On Wild Animals*, Pammenes says that in Egypt live scorpions with wings and two stingers. (He says that he has seen them for himself and is not just repeating what he has heard.) He also says that there are two-headed **SNAKES**—and with two feet where a tail should be. Ctesias says that in Sittace, in Persia, the Argades River is full of snakes with a black body but a white head. They grow to a length of six feet. By day these snakes cannot be seen, for they swim in the water, and at night they come up and kill anyone who approaches the river to wash clothes or draw water. Their victims are countless, because they need water to drink, or else they were too busy during the day to wash their clothes.

{ BOOK XVII }

In his book on India, Cleitarchus says there are **SNAKES** there that are twenty-five feet long. He also says there is a snake that is much different in appearance: it is short, and it looks as if it had been painted, with bronze, silver, red, or gold stripes that run from the head to the tail. He says that this snake has a terrible bite that kills very quickly.

In his *History of the Ptolemies*, Nymphis says that in the country of the Troglodytes there are **VIPERS** that measure about twenty-five feet long, much bigger than other vipers. The tortoises there have shells so big that it would take seventy-two gallons of water to fill them.

In his twelfth book, Phylarchus writes the following about the **ASPS** of Egypt. They are treated with deep respect, and so they are gentle and, in fact, tame. They eat alongside the children and do them no harm, and they come out of their holes when called. The way to call them is to snap your fingers. When the snakes appear, the Egyptians give them special food as a bond of friendship, such as barley wetted in wine and honey. They place this treat on their own supper tables, and then they snap their fingers to call their guests. The asps slither out of their various holes and hiding places, circling the table and raising their heads to lick up the food, eating it all up. If an Egyptian is drawn out of bed by a call of nature in the middle of the night, he snaps his fingers so that the asps will get out of the way. The snakes somehow have learned to distinguish the snap that means to come for a meal and the snap that means to move aside, and so the man who has to get up at night runs no risk.

The **CROCODILE** can grow to extraordinary size. In the reign of Psammetichus, the king of Egypt, a crocodile reached the length of thirty-five feet, and in the reign of Amasis in the following century, there was one that was three or four feet longer than that. In the Gulf of Laconia, I hear, there are huge sea serpents, and Homer writes about these "sea monsters of Lacedaemon." Around Cythera the sea serpents are even larger. Their sinews are good for making bowstrings, catapult bands, and so forth. In the Red Sea, some fish can get to be eight or nine feet long. Amometus says that in Libya the priests can call up crocodiles nearly fifty feet long. Theocles, in his fourth book, says that the sea serpents around Syrtis are longer than a trireme. Onesicretus and Orthagoras say that on the coast of Gedrosia, a huge part of India, the sea monsters are six hundred feet long. When they blow air out of their noses, they churn up tall waves that the ignorant and inexperienced mistake for waterspouts.

Aristotle tells us, in the eighth book of *History of Animals*, that **ELEPHANTS** eat more than a hundred gallons of barley a day, and seventy to eighty gallons of groats, and drink more than a hundred gallons of water in the morning and another eighty or so in the afternoon. He says that they live for two hundred and even three hundred years.

The **CAMEL** abhors clean water and much prefers drinking muddy, filthy water. It will not drink clear water if it comes to it, but steps in and stirs up the mud. It can go for as many as eight days without drinking.

There is a creature called the **ONOCENTAUR**, and if you were to see one, you would know that centaurs existed, made of a blend of different creatures. Whether they came about naturally or some strange miracle of nature melded the body of a horse and a man and gave them one soul, we have no idea. I have heard it said that the onocentaur has the face of a man, but surrounded by a mat of hair. Its neck and upper body are human, but with swollen breasts. Down below, though, its body looks like a donkey's, white in color. Its hands have two purposes, for it moves itself along by means of its hands and feet, going as fast as any four-legged creature. And if it needs to pick something up, or put something down, or grab hold of something and clutch it, it can do so with its feet, which become hands, no longer walking but sitting down to do so. It has a violent disposition. If it is captured, it refuses to eat; it cannot stand imprisonment and would rather die.

Boeotia has no **MOLES**. They do not enter from the neighboring province of Lebadeia, and if one arrives by accident it dies.

Libya has no wild boars or horses. In the Black Sea there are no cephalopods or **SHELLFISH**, or at least only a few of them. Dinon says that in Ethiopia there are birds with a single horn, pigs with four horns, and sheep that have hair like a camel's.

There is a kind of **TOAD** that is lethal to drink and dangerous to look at. If someone crushes this toad and mixes it with wine or some other drink, then gives it to someone else to consume, the effect is instantaneous. To gaze at the toad is dangerous, too. If a man looks at this toad, it will stare back, locking its eyes on the man and breathing on him; the man will turn pale, and anyone who meets him will think that he has been sick for a long time. This pallor lasts only a few days, though, and then goes away.

I do not believe him, but others might when they learn from Eudoxus that he saw **BIRDS** bigger than oxen after passing through the Pillars of Hercules. I do not believe him, as I say, but that does not mean that I will suppress his reports.

Theopompus says that in the third plowing and sowing, the Veneti, who live on the shore of the Adriatic Sea, give presents to the **JACKDAWS**, such as fine cakes made of ground barley, honey, and oil. These presents are meant to persuade the jackdaws to declare a truce and not dig up and make off with the seeds. Lycus adds that the Veneti lay out scarlet ribbons, and the great flocks of jackdaws respect these as boundaries and remain outside them while two or three, acting as ambassadors from different cities, so to speak, go have a look at the presents. After they do, they call out to the other birds, and the flocks descend on the cakes and eat them. Once this has happened, the Veneti know that the truce is on. If the birds decide that the cakes are too thin or not well made enough, then it is war, and the result is famine, for the jackdaws will lay ruin to any field, digging up and making off with any seeds that have been planted.

In his book *Stages*, Amyntas says that in the area around the Caspian Sea there are huge, endless, innumerable herds of cattle and horses. He adds that at certain times of the year hordes of **RATS** descend on the land, and even though the rivers are swift, the rats have no difficulty or hesitation in swimming across them, grabbing hold of one another's tails and moving as a single chain. When they reach the shore, they sweep across the fields and clean them of grain, remove every fruit from the trees, and even eat the branches. The Caspii, to protect themselves, take pains not to kill birds of prey, and these dart down from the clouds and seize the rats, thus helping the Caspii stave off certain famine.

The **FOXES** in the Caspian region are innumerable, too, and are so bold that they actually come into town, acting just like dogs, wagging their tails. The rats are as big as the mongooses of Egypt. They are wild and fierce, and their teeth are so strong that they can bite through iron. The rats of Teredon are about the same size, and the traders there sell their skins to the Persians, for they make fine, warm tunics.

Eudoxus says—believe him if you want to—that the eastern Galatians pray and make certain kinds of sacrifices when **LOCUSTS** descend on their fields. The birds then come and destroy the locusts en masse. If a Galatian captures a bird during this time, then, according to their law, that man is put to death. If a judge shows leniency and pardons the man, then the birds will refuse to respond the next time they are summoned so.

In Libya, they say, there was once a race of people called the Nomaei. They were prosperous and lived in good, fertile country, but they were obliterated when a huge army of **LIONS** invaded. Every last one of the Nomaei men was killed. A massed attack by lions is something no living creature can stand up to.

In the land of the Caspii, I hear, there is a huge lake full of great fishes called **OXYRHYNCHI**. The Caspii chase them down and catch them, salting and brining them and then shipping them by camel caravan to Ecbatana. The oil is extremely rich, and the Caspii anoint themselves with it. They also make glue from it, and this bonds anything it touches. Even if you soak something thus glued in water for ten days or more, it will not loosen. Artists find it particularly useful in making ivory sculptures.

In his *History of Crete*, Antenor says that the gods once sent a swarm of copper-colored **BEES** to attack the city of Rhaucus, stinging the inhabitants terribly. The Rhaucii could not stand up to the attack, so they abandoned the city, but founded another one elsewhere that, out of love for their mother city, as the Cretans say, they also called Rhaucus. Antenor says that descendants of these bees still live on Mount Ida. There are not many of them, but their sting remains fierce.

LIONS love to eat camel flesh, as Herodotus tells us when he writes of the lions that attacked King Xerxes's supply train. They did no harm to the humans or other animals, but devoured the camels. Herodotus shows little knowledge of the food habits of Thracian lions. The Arabians know all about this sort of thing, however, since their country is both the mother and the nurse of all lions. I would not be surprised if this love of camels is inborn, for nature so works that creatures hunger for foods that they have never before encountered.

In India, near the Astaboras River, lies the country of the people they call Root Eaters. In summer, great clouds of **MOSQUITOES** rise up, enough to fill the whole sky, and they cause tremendous suffering. Nearby, in Lake Aoratia, the mosquitoes are particularly abundant. This country is a desert, although, the Indians who live there say, it was not always so. First, scorpions overran the place, and then hordes of four-jawed spiders. The inhabitants fought the foul invaders, but after a while all the men were killed, and the surviving women and children had to flee, their once generous fatherland now become a desert. Perhaps I would not be wrong to say that the place was not even a motherland to them.

Once an invasion of field **MICE** drove a whole tribe of people in Italy from their native country, making them exiles by depriving them of all their food, as a drought or killing frost might. An army of sparrows invaded Media, eating all the seeds and routing the people. A rain of frogs fell from the sky onto the land of the Autariatae, in Mysia, and they had to leave. I have already mentioned the members of a Libyan tribe who were invaded by lions and had to leave their native land.

It would be tiresome and useless to describe the **RHINOCEROS**, since most Greeks and Romans know about the creature and many have even seen it. But it is still worth mentioning some of its characteristics. The tip of its horn is very sharp and is as strong as iron. It sharpens this tip on rocks and then goes off to attack elephants, which otherwise it would have no chance against given the elephants' size and strength. The rhinoceros will come up under an elephant's belly and rip it open, and the elephant soon collapses, drained of blood. In their country, you can see many bodies of elephants that have died in this way. If the rhinoceros is slow, however, the elephant will lasso it with its trunk and then cut the rhinoceros to bits with its tusks. The rhinoceros's skin is so thick that arrows cannot pierce it, but its enemy is very powerful.

In his book *On Europe*, Mnaseas describes a temple dedicated to Heracles and his wife, the daughter of Hera. Around the temple live huge flocks of tame birds that are cared for at the public expense. The **HENS** feed in the temple of Hebe, while the cocks feed in the temple of Heracles. A perennial stream flows between the two temples. Now, no hen ever appears in the temple of Heracles, and no cock ever appears in the temple of Hebe. At mating time, though, the cocks will fly across the stream and couple with the hens, then fly back. When the chicks have hatched, the cocks raise the male birds, and the hens raise the females.

{ EPILOGUE }

I have tried to set out, to the best of my ability, everything that I have been able to learn—painstakingly, through considerable research and hard work—about the matters contained here. If I have omitted anything, it is not out of disrespect for the animals, as if they were beneath my dignity and not worthy of notice. I love knowledge; it is my nature. I know that there are those who think only about earning money, and who will scorn me for not having devoted myself to attaining wealth at the palace instead of spending my time on this work.

I would rather spend time with the foxes, lizards, bugs, snakes, and lions, with the ways of the leopard, the passions of the stork, the sweet song of the nightingale, the wisdom of the elephant, with the myriad multiform fishes and migrating flocks of cranes and the abundant serpents, all the creatures that figure in the book that I have assembled. I could not care less about being counted among those rich men, for I would prefer to be thought of as among the poets, the scholars who are good at puzzling out the workings of nature, the writers—well, I would like to be thought of as wise, or at least knowledgeable about many things, even expert. But enough of that.

I know that some people will object that I have not grouped all the stories about any given animal in one place, but have mixed the stories, writing one thing here and another there. Let me say that I will not bind myself in slavery to what someone else thinks I should do; I do not have to follow another man's lead. Moreover, I wanted to attract readers with a pleasing variety of stories, and not bore them with a monotony of facts. I have therefore tried to weave a story that is like a field of wildflowers of many colors, with each animal being a different kind of flower. Hunters may say that finding an animal involves considerable skill and luck, but I say that there is nothing special in following tracks or making a kill. Learning about the nature of these animals, though— that is something special indeed.

Cephalus and Hippolytus, those great hunters, or Metrodorus of Byzantium and his son Leonidas, or Demostratus, or all those other great fishermen (and there are many of them), might have something else to say about that. The same with the artists— Aglaophon, proud of his portraits of horses, and Apelles of fawns, and Myron of his statue of a calf. I have searched out and brought forth everything I could find out about animals—about their ways, shapes, wisdom, cleverness, righteousness, courage, sobriety, and sense of filial duty. None of this will earn me instant recognition. At this point in the narrative, I am instead saddened to note that on that last matter, filial piety, animals without reason are more to be praised than us humans. I won't go on about that. As I said at the beginning, I should not be criticized for repeating what many writers before me have said about the animals. I cannot create those animals for myself, after all, though I think it is plain that I have come to know about a lot of them.

Indeed, I have written more about them, and provided more new information, than any other writer. I cherish the truth, and readers who approach this book with an open mind will soon appreciate the quality of my book, the hard work I have put into it, the elegance of its style and structure, and the appropriateness of the words and phrases I have used.

{ NOTES }

4 The quotation from Gerald of Wales is from *The History and Topography of Ireland* (Penguin, 1982), 51.

8 The island in the Adriatic Sea is probably San Domino, one of the Tremiti Islands, now part of Italy's Gargano National Park. The bird is probably Cory's shearwater (*Calonectris diomedea*), a gregarious species that follows fishing ships in the Mediterranean to this day. See D'Arcy Wentworth Thompson, "The Birds of Diomede," *Classical Review* 32–33 (February–March 1918): 92–96. See also John Pollard, *Birds in Greek Life and Myth* (Thames & Hudson, 1970).

9 Glauce was a famed harpist from Chios, and rumored to have been the mistress of Ptolemy II. See Roger French, *Ancient Natural History* (Routledge, 1994), 183, for more on this story and its reception. Soli, near the present Turkish city of Mersin, is known to speakers of English today only for a curious etymology. The colonists there spoke a dialect of Attic Greek, but badly—so much so that Soli became a byword for a backwater where the mother tongue and other accouterments of civilization were likely to be mangled and maltreated. Thus our word *solecism*.

"A bee is never caught in a shower," the English proverb has it. "In times of heavy wind": This report is also found in Pliny and other sources. Roger French observes in *Ancient Natural History* (Routledge, 1994) that the pebble-ballast notion is "a very common story in antiquity" (205).

10 Pliny shares Aelian's dislike for rays; see the notes by volume editors

John Bostock and H. T. Riley in *The Natural History of Pliny* (George Bell & Sons, 1890), 2:411.

This Nicias is not to be confused with the Athenian general of Peloponnesian War renown.

11 Aelian's readers would have known the myth of Tithonus, a handsome young man who became the lover of Eos, the goddess of the dawn. She granted him the gift of immortality, but Tithonus aged all the same, becoming smaller and smaller and his voice creakier and creakier. Out of pity, Eos turned Tithonus into a cicada, whose song the ancient Greeks and Romans considered beautiful. See Rory B. Egan, "Cicada in Ancient Greece," in the online cultural entomological publication *Bugbios* (www. insects.org/ced3/cicada_ancgrcult.html), accessed Dec. 10, 2010.

Ergane, "the doer," is an aspect of the goddess Athena. The relevant fable of Athena and Arachne is found in Ovid's *Metamorphoses*.

12 For more on the accursed avengers Orestes and Alcmeon, see the plays by Aeschylus, the *Oresteia* and *Seven Against Thebes*.

In Greek mythology, Caeneus was a warrior who began life as a girl whom Poseidon, the sea god, raped. She begged to be turned into a man so that she could never be raped again, and Poseidon granted her wish. After he died in the legendary battle of the Lapiths and the centaurs, Caeneus became female once more. Versions of the story are found in Virgil's *Aeneid* and Ovid's *Metamorphoses*. Teiresias was transformed into a woman for seven years after having struck a pair of copulating snakes with his staff; the snakes were sacred to Hera, who, according to Hesiod, punished him accordingly.

13 "Homer says": *Iliad* 20.321.

"Numbfish": Aelian is making a play on words on the name of the fish called, in English, the torpedo, a kind of electric ray. In Greek, it is *narké*, related to the word meaning a sleep so deep it approaches a coma. The English play on words "torpedo"/"torpid" approaches but does not quite get at this deep numbness.

"squill leaves": Writes Maude Grieve, the botanist and author of *A Modern Herbal*, "Merck, in 1879, separated the three bitter glucosidal substances Scillitoxin, Scillipicrin and Scillin. The first two are amorphous and act upon the heart, the former being the more active; Scillin is crystalline and causes numbness and vomiting. Other constituents are mucilaginous and saccharine matter, including a peculiar mucilaginous carbohydrate named Sinistrin, an Inulin-like substance, which yields Laevulose on being boiled with dilute acid. The name Sinistrin (in 1834, first proposed by Macquart for Inulin) has also been applied to a mucilaginous matter extracted from barley, but it remains to be proved that the latter is identical with the Sinistrin of Squill. Calcium oxalate is also present, in bundles of long, acicular crystals, which easily penetrate the skin when the bulbs are handled, and causes intense irritation, sometimes eruption, if a piece of fresh Squill is rubbed on the skin."

13–14 The lousefly, *Stenopteryx hirundinis*, is a known parasite in swallow nests. The original Greek gives the word *silphé*, whose usual meaning is "cockroach."

15 Birdsong, canine barks, and every other sound that an animal can produce with its mouth are the result of neural circuitry established hundreds of millions of years ago by fish, which communicate by humming, grunting, and other vocal signs. Small wonder, then, that the stingrays should be enchanted. See A. H. Bass, E. Gilland, and R. Baker, "Evolutionary Origins for Social Vocalization in a Vertebrate Hindbrain-Spinal Compartment," *Science* 321 (July 18, 2008): 417–421.

"Homer tells of this fact": *Iliad* 17.674. In the sonorous neo-Elizabethan prose of A. T. Murray, "fair-haired Menelaus departed, glancing warily on every side as an eagle, which, men say, hath the keenest sight of all winged things under heaven, of whom, though he be on high, the swift-footed hare is not unseen as he croucheth beneath a leafy bush, but the eagle swoopeth upon him and forthwith seizeth him, and robbeth him of life."

16 "as Homer tells us": *Iliad* 14.233.

Carl Gustav Jung observes that the myth of the punished raven survives in German folklore: "It is said that the raven has to suffer from thirst in June or August, the reason being that he alone did not mourn at the death of Christ, and that he failed to return when Noah sent him forth from the ark." C. G. Jung, *Four Archetypes* (Routledge and Kegan Paul, 1972), 429.

17 Stingrays are in fact responsible for several human deaths each year, most famously that of the Australian television personality Steve Irwin, "the Crocodile Hunter," in September 2006. Aelian alludes here to Achilles's "famous spear of ash" (*Iliad* 16.143).

18 Ancient Greek and Roman literature is full of references to the cerastes. The various small, venomous vipers of that genus extend from North Africa across the Arabian Peninsula into Iran, territory well known to merchants, soldiers, and travelers in Aelian's time. Even so, the qualities ascribed to them are often legendary, a penchant that survives in medieval bestiaries and beyond. Here is Leonardo Da Vinci: "This has four movable little horns; so, when it wants to feed, it hides under leaves all of its body except these little horns which, as they move, seem to the birds to be some small worms at play. Then they immediately swoop down to pick them and the Cerastes suddenly twines round them and encircles and devours them" (*Notebooks* 1256).

"malmignatte": *Latrodectus tredecimguttatus*, the European black widow. Herodotus (4.173) tells us that the Psylli went extinct after a sandstorm blew up and buried them.

19 Mallow and spurge contain alkaloids that are poisonous to a broad range of animals and to humans as well. Mulleins contain high levels of coumarin, a blood-thinning agent widely used as rat poison. In the First Sacred War (595–585 BCE), the Amphictyonic League poisoned the water supply of the Greek city of Kirrha with hellebore, the first known instance of chemical warfare in human history. See Adrienne Mayor, *Greek Fire, Poison Arrows and Scorpion Bombs: Biological and Chemical Warfare in the Ancient World* (Overlook Press, 2003), and Borden Institute / U.S. Army Medical Department, *Medical Aspects of Chemical Warfare* (Office of the Surgeon General, 2008).

20 Identified as the Maritza River in present-day Bulgaria, the Hebrus was esteemed as among the most scenic of rivers in the ancient world. Writes Alcaeus, "Hebrus, you flow, the most beautiful of rivers, past Ainos into the turbid sea, surging through the land of Thrace . . . and many maidens visit you to bathe their lovely thighs with tender hands, enchanted as they touch your soft waters" (frag. 45A).

21 Aelian is likely writing here of the fruit fly, *Drosophila melangaster*.

Pliny elaborates: "Once a basilisk was killed with a spear by a man on horseback; the venom passing up through the spear killed not only the rider but the horse as well." Spanish travelers in North America identified the Gila monster (*Heloderma suspectum*) as a basilisk, but the chroniclers of the ancient world depict the basilisk as a small snake or a monster half-snake, half-rooster.

21–22 Poroselene has been identified as modern Hekatonisa (Greek) or Cunda (Turkish), a small island between Lesbos and the Turkish mainland.

24 Germanicus Caesar is better known to us as Caligula.

25 Xenophon records, among other things, that Shah Darius had a thousand animals slaughtered each day for the royal table at his capital of Persepolis. For more on the intelligence and emotions of elephants and other creatures, see Jeffrey Moussaieff Masson and Susan McCarthy, *When Elephants Weep: The Emotional Lives of Animals* (Delta, 1996).

26 The Rhyndacus flows across northwestern Turkey into the Black Sea, crossing the ancient provinces of Mysia and Phrygia. Writes Pliny the Elder, "Megasthenes informs us that in India, serpents grow to such an immense size, as to swallow stags and bulls; while Metrodorus says that about the river Rhyndacus, in Pontus, they seize and swallow the birds that are flying above them, however high and however rapid their flight."

29 The shrewmouse fascinated Buffon as well: "The shrewmouse seems to form a link in the chain of small animals, and to fill the interval between the rat and the mole, which, though they resemble each other in size, differ greatly in figure, and are very distant species. The shrew is still smaller than the mouse, and has an affinity to the mole, by its long

nose; by its eyes, which, though larger than those of the mole, are much concealed, and more minute than those of the mouse; by the number of its toes, having five on each foot; by the tail and legs, especially the hind-legs, which are shorter than those of the mouse; by the ears; and, lastly, by the teeth. This little animal has a strong and peculiar odour, which is very disagreeable to the cats, who pursue and kill, but never eat the shrews. It is probably this bad smell, and the reluctance of the cats, which have given rise to the vulgar prejudice, that the bite of the shrew-mouse is venomous, and particularly hurtful to horses. But the shrew is neither venomous, nor is it capable of biting; for the aperture of its mouth is not large enough to take in a duplicature of another animal's skin, which is absolutely necessary to the action of biting. The disease of horses, vulgarly ascribed to the bite of the shrewmouse, is a swelling or blotch, and proceeds from an internal cause, which has no relations to a bite." Georges Louis LeClerc, Comte de Buffon, *Natural History: General and Particular*, trans. William Smellie (T. Cadell, 1781), 4:305.

30 "Aristotle says": see *History of Animals* 618 b 11.

31 "wild lettuce": This bit of folk wisdom has basis in scientific fact. Foods rich in the carotenoid lutein, including lettuce, have proven value in combating macular degenerative diseases and other eye maladies.

33 The province of Moesia lay in what is now northern Serbia and southwestern Bulgaria, bounded by the Balkans and the Danube River. The Scythia of Aelian's day lay to the east, in present-day Ukraine. Aelian refers to Herodotus, who writes (5.10) that he heard from Thracian sources that bees live no farther north than the Danube (Ister). "I think it is likely," he adds, "that the cold renders the northern countries uninhabitable."

35 Carmania was a satrapy in Persia, and in Aelian's time a byword for the remotest end of the earth. Molossia, on the border of present-day Albania and northwestern Greece, was renowned for its fierce guard dogs. Writes Virgil, "Never, with them on guard, need you fear a mid-night thief in your stalls, or an attack by wolves, or an ambush by Iberian bandits" (*Georgics* 3.405), while Horace calls them "the shepherd's friend" (*Epodes* 6.5).

37 The trochilus is probably the Egyptian plover (*Pluvianus aegyptius*), neither a true plover nor now resident in Egypt. The Egyptian plover is informally called the "crocodile bird" to honor this symbiotic relationship, which ornithologists have remarked on anecdotally but have not yet fully documented scientifically; the reports are that the bird eats rotting meat from the crocodile's back teeth, a useful bit of dentistry. Herodotus tells the same story.

38 "to quote Euripides": *Ion* 1198.

"Euripides says": The theme of jealousy as curse runs through *Andromache*, *Medea*, and *Hippolytus*.

"The mare tries to keep this secret": Lucan records this belief in the *Pharsalia*, while Aristotle writes in *History of Animals*, "After parturition the mare at once swallows the afterbirth, and bites off the growth, called the hippomanes, that is found on the forehead of the foal. This growth is somewhat smaller than a dried fig; and in shape is broad and round, and in color black. If any bystander gets possession of it before the mare, and the mare gets a smell of it, she goes wild and frantic at the smell. Vendors of drugs and potions covet the substance."

"If this is so": Presumably, the seal would have to be killed in order to gain access to this milk, so from the seal's point of view this seems perfectly defensible behavior. A nineteenth-century medical textbook describes a patient's being given "internally curdled milk" following an epileptic seizure; see Thomas Alexander Wise, *Review of the History of Medicine* (Oxford University Press, 1867), 2:175.

39 Eudemus of Rhodes (ca. 370–300 BCE) was a philosopher and historian of science whose work was overshadowed by his contemporary Theophrastus. Both were students of Aristotle's and edited his works.

40 Alexander of Myndus, who probably lived a generation or two before Aelian, was the author of *A History of Beasts* and *On Birds*, among other treatises. His works were highly influential in antiquity, but only fragments have survived.

41 The pappus is unidentified. In modern Greek, *pappos* refers to any kind of small bird, what American ornithologists informally call LBJs:

"little brown jobs." Sirius rises in mid-July. The cuckoo does not go into hiding, but it moves on; breed-parasitic cuckoos time their migrations to the nesting patterns of their hosts. See Robert B. Payne, *The Cuckoos* (Oxford University Press, 2005), 24.

43 "Aphrodite sets out during this time": this may allude to an astronomical event, for Aphrodite is, of course, the Greek counterpart of Venus. As to pigeons more specifically, the turtledove and rock dove were sacred to Aphrodite, though their kin the wood pigeon seems not to have been. See Leonard Whibley, *A Companion to Greek Studies* (Cambridge University Press, 1905), 30. The reference to Homer is at *Iliad* 5.427.

44 "Three thousand mares": *Iliad* 20.221. "Boreas fell in love": *Iliad* 20.223. Boreas is the minor deity of the north wind. For Aristotle, see *History of Animals* 572 a 16.

45 The stone the Greeks called "lynx piss" was probably tourmaline. Theophrastus wrote of its power to attract straw and small bits of wood, the process called pyroelectricity.

"none gathers": Indeed, birds of prey tend to be solitary, a notable exception being the Harris' hawk (*Parabuteo unicinctus*). All of them, however, imbibe and expel water.

46 "Ctesias says": The *Suda*, a Byzantine encyclopedia, records of Ctesias of Cnidus, "He was the son of Ctesiarchus or Ctesiochus, from Cnidus. As a physician, he cared in Persia for [king] Artaxerxes II Mnemon, who had ordered him to come. He composed *History of the Persians* in twenty-four books."

"The trainers keep the meat": This is one of the first mentions of falconry in ancient literature. Some sort of falconry was practiced in northern Greece and Macedonia, but it was uncommon in Rome until late imperial times, attested by a floor mosaic depicting a trained hawk hunting ducks.

47 Homer writes (*Odyssey* 7.291) that Argos waited faithfully for twenty years for his master to return from the Trojan War and his subsequent *periplus*.

48 The Dionysia, held in March and April, and the Lenaea, held in

January, were festivals dedicated to the wine god Dionysus. The Festival of the Urns commemorated the dead. Causeway Day, also in March, seems to have been an exercise in hurling vulgarities at fellow walkers along the Sacred Way between Athens and Eleusis.

"The Indians crush them": Aelian is describing the lac insect, *Kerria lacca*. *Cynocephalus* is used as a generic term for simians; Aelian often uses it to mean baboons. It is anyone's guess what he means here.

49 Monkshood, also called aconite or wolfsbane, is a member of the buttercup family whose flowers can be toxic if eaten in sufficient quantity. It figures in the traditional formulary of several Asian traditions as a diuretic and purgative—but also as an arrow poison.

The cyan bird may be the western rock nuthatch (*Sitta neumayer*), a shy, blue bird endemic to the eastern Mediterranean and given to making its nests among rocks. Skryos, the southernmost island in the Sporades, is still lightly populated; in antiquity it was a byword for the boondocks, famed chiefly as the place where Theseus died.

50 Aelian refers to the myth of Glaucus, which Euripides mentions in the *Alcestis*. See Leonard Muellner, "Glaucus Redivivus," *Harvard Studies in Classical Philology* 98 (1998): 1–30.

51 Ctesias writes in his *Indica*, "In the river Indos a worm is found resembling those which are usually found on fig trees. Its average length is seven cubits, though some are longer, others shorter. It is so thick that a child ten years old could hardly put his arms round it. It has two teeth, one in the upper and one in the lower jaw. Everything it seizes with these teeth it devours. By day it remains in the mud of the river, but at night it comes out, seizes whatever it comes across, whether ox or camel, drags it into the river, and devours it all except the intestines. It is caught with a large hook baited with a lamb or kid attached by iron chains. After it has been caught, it is hung up for thirty days with vessels placed underneath, into which as much oil from the body drips as would fill ten Attic kotylae [i.e., a little more than a gallon]. At the end of the thirty days, the worm is thrown away, the vessels of oil are sealed and taken as a present to the king of India, who alone is allowed to use it. This oil sets everything

alight—wood or animals—over which it is poured, and the flame can only be extinguished by throwing a quantity of thick mud on it."

53 Astypalaea and Rhenea are small islands in the Cyclades chain. None of Aristotle's extant works mentions the islands or the qualities Aelian mentions here. As for Elis, Herodotus (4.30) writes that the residents of the southern Italian town had to drive their jacks and jennies to another town to get them to mate.

Modern Reggio Calabria and Locri, in southern Italy, lie some forty miles apart and are separated by several rivers; it is not clear which one Aelian means. Pausanias, who tells a version of the story, identifies it as the Caecinus, perhaps the modern Marro River. See Ettore Pais, *Ancient Italy: Historical and Geographical Investigations in Central Italy, Magna Graecia, Sicily, and Sardinia* (University of Chicago Press, 1907), 50–51. Cephallenia, the modern Kefallinia, is the largest of the Ionian islands.

Aelian probably means the Pseudo-Aristotle, to whom the *Mirabilia* is attributed. Theophrastus tells the same story, but in his version, the metal is gold. See William W. Fortenbaugh, *Theophrastus of Eresus: Sources for His Life, Writings, Thought, and Influence* (E. J. Brill, 1992), 66. Amyntas was a bematist who recorded in his book *Strathmoi* (Stages) Alexander the Great's travels. See page 152.

54 "the people dispersed": *Iliad* 24.1.

55 The purple gallinule, *Porphyrio porphyrio*, also called the purple swamphen or purple coot, is a rail of the Mediterranean region. The bird is gregarious and quite noisy, for which, we learn from Pliny, the Romans kept gallinules as watchbirds, as they did peacocks. So valuable was the gallinule in this service that the Romans, omnivorous and wide-ranging in taste, apparently did not make the bird part of their larder.

56 Best known today for theorizing about the existence of atoms, Democritus (ca. 460–370 BCE) was the author of *Causes Concerning Animals*, a work, now lost to us, that was influential in antiquity. "The tireless sun": *Iliad* 18.239. "Like cattle frightened by a lion": *Iliad* 11.172.

57 "Aristotle says": *History of Animals* 552 b 20. The Hypanis is the Bug River of southern Ukraine. Its name has nothing to do with the fly.

Marjoram, a natural disinfectant, has antifungal and antibacterial properties. It figured in Greek and Roman medicine, despite Aelian's discounting it here.

58 The Sacae, or Saka, were a Central Asian people, probably a Scythian tribe who lived on the western extremes of what is now Xinjiang, the desert province of China. Pliny describes them in *Natural History*, though only his *Sacae tigraxauda*, "Sacae with pointy hats," were true Saka.

60 Crocodile City, or Crocodilopolis, lay on the west bank of the Nile River near Faiyum, south of present-day Cairo. Mares and Moeris are variants of the Greek name given to the Pharaoh Amenemhet III (r. 1831–1786 BCE).

61–62 Aristotle, it is worth noting here, was fascinated by hyenas, which he had presumed to be hermaphroditic and capable of asexual reproduction. After dissecting a male spotted hyena, he revised his view, though he might have been more confused had the subject been female, since female spotted hyenas "are the most highly 'masculinized' extant female mammals." The quoted words are from Stephen E. Glickman, Gerald R. Cunha, Christine M. Drea, Alan J. Conley, and Ned J. Place, "Mammalian Sexual Differentiation: Lessons from the Spotted Hyena," *Trends in Endocrinology and Metabolism* 17, no. 9 (2006): 349–356.

63–64 This method of scorpion control was traditionally used in Mexico and the American Southwest as well.

64 "The fox knows many things, but the hedgehog knows one big thing," Archilochos famously wrote. Aesop also tells stories of the enmity between the fox and the hedgehog.

65 The belief that beavers engaged in autocastration figures in several medieval bestiaries, often with a gloss of this sort: "If a man wishes to live chastely he must cut off all his vices and throw them from him into the face of the devil. The devil, seeing that the man has nothing belonging to him, will leave the man alone."

66 The dogfly is probably the stable fly, *Stomoxys calcitrans*, a significant pest to domesticated livestock.

67 "Aristotle says": *History of Animals* 577 b 30. Work on the Parthenon

began in 447 BCE and ended in 438 BCE. The Prytaneum was the Athenian equivalent of city hall, housing the office of the chief magistrate and the community's altar or hearth. Ambassadors, distinguished foreigners, and honored citizens were fed there, and so were prized athletes.

68 *Conopeum* was the Latin form of the Greek word for gnat, *konopeion*, which begat the English word *canopy*, meaning a netting to keep insects away from sleepers. Thus, it would appear, "Gnat City," somewhere in the vicinity of modern-day Yalta. The Maoetic Sea is the Sea of Azov, a northerly branch of the Black Sea.

69 Susa, in modern Iran, was a capital of the Elamite Empire. The Roman emperor Trajan conquered it in 120 CE, just a couple of generations before Aelian's birth, marking the easternmost limit of the Roman Empire.

70 Paeonia was a province north of Macedonia in what is now southwestern Bulgaria. The "one-eye," *monops*, was probably the aurochs (*Bos primigenius*), an ancestral cow that survived in Europe until the last was killed in Lithuania in 1627.

Many of the books assigned to Aristotle in the Roman era have since been reassigned to the so-called Pseudo-Aristotle or to Theophrastus. In this case, the attribution should be to Theophrastus's *De signis tempestatum*, a treatise on weather signs. See John Scarborough, "Nicander *Theriaca* 811: A Note," *Classical Philology* 75, no. 2 (April 1980): 138–140.

74 Aelian alludes to the laws of the perhaps legendary ruler of Sparta, Lycurgus (ca. 750 BCE), which commanded filial piety, but which also took children away from parents for communal education. The reference is thus a little murky. See Plutarch's *Lives*.

75 Medea and Procne were infamous as mothers who, among other crimes, murdered their own children.

76 "as Homer says": *Iliad* 1.82. Juba, who sided with Pompey and other Roman senators against Julius Caesar and committed suicide in 46 BCE after being defeated, was less renowned than his son, Juba II (50 BCE–24 CE), who wrote many books in Greek on history, geography, grammar, and drama that Aelian doubtless read. On conquering Mauretania, Julius Caesar paraded the five-year-old Juba II in Rome, but then allowed him

to be raised as a Roman citizen. Octavian, who would become Augustus Caesar, befriended the well-educated Juba II and elevated him to the throne of Numidia, a province newly added to the Roman Empire. In 25 BCE he was appointed the ruler of Mauretania, which he governed until his death from his capital city of Caesarea, the modern-day Cherchel, Algeria. On top of all that, he married Cleopatra Selene, the daughter of Mark Antony and Cleopatra, who was renowned for her beauty.

79 "seed weakens over time": This may be the earliest expression of the idea of dominant and recessive genes in the literature. Similar beliefs were current until Gregor Mendel's time.

80 Apart from being one of the wealthiest men in its history, Marcus Licinius Crassus (115–53 BCE) jointly ruled Rome with Julius Caesar and Pompey, having attained renown for suppressing the slave rebellion led by Spartacus in 70 BCE. Crassus's death sparked civil war between the forces of Pompey and Caesar. Domitius Ahenobarbus served with Crassus as censor in 92 BCE.

83 The Antandria and Scamander are rivers in the Troad of Asia Minor. The Greek word *xanthus* means "pale."

"It is called the hunter": Aelian might be describing the myna, introduced to Rome from India, though the myna's preferred food is insects and not small birds.

85 "which supposedly knows nothing of law": Aelian alludes to a fragment of Euripides here.

87 No known eagle is anything but carnivorous. The followers of Pythagoras, the mathematician and philosopher, however, were not allowed to eat meat. To complicate matters, they could not eat beans, either. See the article "Pythagoreanism" in Edward Craig and Luciano Floridi, eds., *Routledge Encyclopedia of Philosophy* (Routledge, 1998), 860.

88 "Theophrastus is our witness": Nothing in the surviving work of Theophrastus supports this, but that is not to say that a lost book did not back Aelian up.

To this day, fennel is used as a folk remedy for poor eyesight in several cultures and traditions.

From other sources, we have evidence that hunters throughout its range poisoned the tips of their arrows with wolfsbane, also called monkshood (*Aconitum napellus*), to kill wolves and other large predators. The plant figures prominently in ancient and medieval herbals. See page 49 and its note.

90 Hippys of Rhegium was the first western Greek historian, the author of an important history of Sicily; see John Marincola, ed., *A Companion to Greek and Roman Historiography* (Blackwell, 2007), chapter 15. The story of the poor worm-besieged woman is known to us only from Aelian. See James Longrigg, "Philosophy and Medicine: Some Early Interactions," *Harvard Studies in Classical Philology* 67 (1963): 147–175.

"as Aristotle notes": *History of Animals* 552 b 1. The caterpillar is probably that of the codling moth, *Carpocapsa pomonella*, a major pest of apple and walnut trees.

93 "Herodotus says": 3.103.

94 "foremost tactician": The allusion is to *Iliad* 2.555.

95 The stone of Heraclea is a magnet.

"for he writes": *Iliad* 15.237.

Manetho, who lived in the third century BCE, was the author of an influential work of history, the *Aegyptica*, a book that remains of great importance to Egyptologists.

96 "Eudoxus says": This is probably Eudoxus of Cyzicus (ca. 130 BCE), a navigator who explored the Red Sea and Arabian Sea. See Henry F. Tozer, *History of Ancient Geography* (Biblo & Tannen, 1964), xxiii, 189–190.

Though it shares its name, the persea tree here is not to be confused with the American persea, or avocado. The Egyptian persea was confined to what are now Egypt, Sudan, and Ethiopia. See C.A. Schroeder, "The Persea Tree of Egypt," *California Avocado Society Yearbook* 61 (1977): 59–63.

Aelian turns another ancient misapprehension slightly sideways here, namely that crocodiles cry while eating. That misapprehension, which gave rise to our expression "crocodile tears," lasted well into the Middle Ages. John Mandeville writes of Egypt, "In that contre . . . ben gret plentee of Cokadrilles. . . . Theise Serpentes slen men, and thei eten hem wepynge"—that is, "In that country there are many crocodiles. These serpents slay people, and then, weeping, eat them."

97 Typho is a Roman name for the god Set, the enemy of Horus or Apollo, for whom the city of Apollinopolis Magna was named. Apollinopolis was famed in ancient times for its enmity to crocodiles and those who worshipped them. Psammenitus ruled for only six months in 526 BCE.

The Vaccaei were a Celtic people who lived in the area of modernday Valladolid, Spain. Diodorus Siculus, the historian, considered them the most refined of the Iberian Celts. As to the fasces, Roman belief held that Romulus appointed twelve officials called lictors to the government of the newly founded city of Rome because he had seen twelve vultures flying over the site. The lictors carried fasces, but they were not royal; instead, they were appointed from the plebeian class or from the ranks of freed slaves. In his histories of early Rome, Livy opines, sensibly, that the Romans borrowed the idea of a dozen lictors from the Etruscans, whose royal government incorporated twelve provincial governors.

97–98 The dogface or doghead has been variously defined. Aelian may be referring here to the hamadryas (*Papio hamadryas*), a baboon native to the Horn of Africa. The dogface may also be the mandrill (*Mandrillus sphinx*), however, which more closely matches Aelian's description but is native to central Africa and not "on the way to Ethiopia."

98 "Wolf light": *Iliad* 7.433. "Wolf-master": *Iliad* 4.101. Archaeologists have not found the statue of the she-wolf at Delphi; see Pausanias, *Description of Greece* 10.14.7.

102 The Daunii were an Italic people of northern Apulia, in the area of Manfredonia and the Gargano Peninsula.

Curias, known in Greek as Kourion, is a promontory on the south coast of Cyprus. The temple to Apollo there was famous throughout the ancient world.

Antigonus Monopthalmus ("one-eyed") was a satrap and general of Alexander the Great's. After Alexander died, Antigonus conquered much of Asia Minor in his own name, and it was ruled for six generations by members of the Antigonid Dynasty. He took Megara, a major city on the Isthmus of Corinth, sometime between 272 and 268 BCE. See N. G. L. Hammond and F. W. Walbank, *A History of Macedonia: 336–167 B.C.* (Oxford University Press, 1988).

104 Once confined to Hellenistic Egypt, the cult of the god Serapis was popular in Rome in Aelian's day.

105 The ruins of Myra, the capital of ancient Lycia, lie about a mile from present-day Demre, in southwestern Turkey.

Bocchoris, or Bakenranef, was a pharaoh famed for his wisdom. He ruled as sole king of the 24th dynasty (ca. 722–ca. 715 BCE) until a rival pharaoh, Shabaka, founded the rival 25th dynasty and conquered Egypt, ordering Bocchoris to be burned alive after he did so.

106 Aelian's animosity toward martens—blameless creatures, except around a chicken coop—was widely shared until very modern times. Today they are esteemed, at least in England, for helping control the spread of the invasive gray squirrel, an import from America.

107 The Athenian lawgiver Phocion (ca. 402–ca. 318 BCE), called "the Good," was judicially murdered by members of the aristocracy whom he had opposed. See Plutarch's *Lives*.

108 "they named a city after them": Leontopolis, near modern Tell al Muqdam in Egypt. Located in the Nile Delta, the city contained temples dedicated to the lion goddesses Bast and Sekhmet.

110 The Strymon is the Struma River of modern Bulgaria and Greece, whose name comes from the same Indo-European root as our word "stream." The sheatfish is a freshwater catfish found in Europe.

111 On the trochilus, see page 37 and its note.

113 "the baby was called Gilgamos": Aelian here relates a detail from the life of Gilgamesh, the Babylonian demigod.

The Achaemenid Dynasty ruled Persia from 705 to about 330 BCE, when Alexander the Great overthrew it. The Achaemenid Empire was the largest in ancient history, extending from India to Europe and North Africa.

114 "a kind of cartilaginous thing": Aelian may be referring to the scalloped torpedo ray (*Torpedo sinuspersici*) or starry skate (*Raja radiata*), though why he should be bringing monkeys into the discussion is a curiosity.

116 In 390 BCE a Gallic army led by Brennus swept down from northern Italy and defeated a Roman army at the Battle of the Allia. Brennus then entered Rome and sacked the city for seven months, finally leaving after failing to take the Capitoline Hill. For complex reasons, Marcus Manlius was later executed at the order of the Roman Senate.

117 "The people of Eretria": Amarynthos, settled since Mycenean times, is in Euboea, in coastal Greece.

118 The Crathis, also called the Sybaris, is the modern Crati River in Calabria, in southern Italy. The people of Croton—who despised the Sybarites "as Bostonians despise Chicago," to quote a *New York Times* article about an 1881 excavation there—redirected the river into the heart of the enemy city, burying it in silt. See Timothy W. Potter, *Roman Italy* (University of California Press, 1987), 23–24.

119 Lying to the northwest of the Acropolis, the Kerameikos district was famed in antiquity for both its potteries (whence our word "ceramic") and its vast necropolis.

Kyrnos (Latin Cyrnus) was the Greek name for Corsica.

120 "shepherd's wedding songs": Euripides, *Alcestis* 577.

For more on Arion, see Herodotus 1.23–24.

123 In another variant, the eagle gave signs to the farmer Gordias that he would become king of Phrygia. Gordias took his oxcart to a nearby village and, naming it Gordian, declared it his capital. The rope that he tied around the cart yoke was the famed Gordian knot. During his kingship, Gordian adopted Midas, he of the golden touch, and made him his heir. See Arrian's *Anabasis* 2.3; see also Lynn E. Roller, "Midas and the Gordian Knot," *Classical Antiquity* 3, no. 2 (October 1984): 256–271.

124 Dikaiarchia is the Latin Puteoli, modern Pozzuoli, which lies near Naples in the province of Campania. Lying in a caldera field, Pozzuoli was reputed in ancient times to be a gateway to the underworld.

125 Prasia has not been identified, but may be the modern Indian state of Kerala. Taxila was an ancient city located in what is now the Punjab province of Pakistan; wild elephants are no longer found in that country.

126 The trochus has yet to be identified, though today the name is that of a genus of sea snails, the largest of which are just six inches or so across. The context suggests that it might be a whale.

For more on the numbfish, see page 13 and its note.

127 The Taulantii were a people of ancient Illyria, in what is now Albania.

128 For more on the lynx, see page 45 and its note. The European lynx is nearly extinct; when it disappears, as it is predicted to do within the twenty-first century, it will be the first big cat to do so since the Pleistocene.

130 The worm beloved of the Indian king may be the red palm weevil, though it seems to be a delicacy no more. See D.S. Hill, *Pests of Crops in Warmer Climates and Their Control* (Springer Verlag, 2008).

The myrus is *Echelus myrus*, the bluntsnout snake eel, described by Linnaeus in 1758. Aelian's story retained currency. Writes the late medieval German alchemist Cornelius Agrippa (1486–1535), "There is a certain fish or great serpent called myrus whose eye, if it be pulled out and bound to the forehead of the patient, is said to cure the inflammation of the eyes, and that the eye of the fish grows again, and that he is taken blind who did not let the fish go." Heinrich Cornelius Agrippa von Nettesheim, *Three Books of Occult Philosophy* (Llewellyn, 1992), 69.

131 "Who bore fifty daughters": *Theogony* 3.240–265; *Iliad* 18.35–107.

134 Claudius Demostratus was a Roman senator who wrote a twenty-six-book treatise on fishing, as well as a book on rivers, that are known to us today through quotations and paraphrases from Pliny, Aelian, and a few other ancient sources. See Michael Petrus Josephus van den Hout, *A Commentary on the Letters of M. Cornelius Fronto* (E. J. Brill, 1999), 278–279.

135 Perimula has been identified as a city at the mouth of the Terengganu River, in modern-day Malaysia. The region, on the South China Sea, is active in pearl fishing to this day.

For more on Juba, see page 76 and its note.

Writes Aelian elsewhere, "The inhabitants of Thebes, although they are Greeks, worship a marten, I hear, and say that it was the nurse of Herakles, or if it was not the nurse, that when Alkmene was in labor and unable to bring her child to birth, the marten ran by her and opened her constricted womb, so that Herakles was delivered and at once began to crawl."

136 Canobus: The story of Canobus is found in several ancient sources, almost all dealing with animals. According to legend, Canobus was buried at the mouth of the Nile northeast of what would become Alexandria; a town that bore his name in antiquity grew up on the site.

137 Cossitinus River: See Herodotus 7.109 on the route of the Persian army of Xerxes, whose magi sacrificed white mares nearby. Diomedes's mares, of course, were made wild by eating human flesh; his eighth labor required Herakles to capture them, which, in one version of the myth, he did by feeding Diomedes to them. Potnaie was the home of Glaucus, who was torn apart by mares. See Robert B. Strassler, ed., *The Landmark Herodotus: The Histories* (Pantheon, 2007), 538–539.

138 In Greek mythology, the Pygmies were a diminutive African tribe that lived along the southern banks of Okeanos, the world-encircling river. For more on the African queen Gerana, see *Iliad* 3.37; Ovid, *Metamorphoses* 6.89–92; see also Joseph Fontenrose, *Python: A Study of Delphic Myth and Its Origins* (University of California Press, 1959), 100–101.

139 In *A Glossary of Greek Birds* (Oxford, 1895), the eminent naturalist D'Arcy Wentworth Thompson identifies this bird as the fruit pigeon, species of which range from West Africa to South Asia.

140 "In India there is another bird": Aelian describes the hill myna (*Gracula religiosa*), renowned for loquaciousness.

For more on classical beliefs about the hoopoe, see Arthur Bernard Cook, *Zeus: A Study in Ancient Religion* (Biblo & Tannen, 1964), 44–45.

141 "Among the Prasii": This is likely the gray or Hanuman langur (*Semnopithecus entellus*), which devout Hindus regard as sacred. The Prasii ("Easterners") lived along the Ganges and were reputed to have

prevented Alexander the Great from crossing the river. See W. W. Tarn, *Alexander the Great* (Cambridge University Press, 2003), 281–282.

142 "grass-eating animal": Aelian is describing the yak (*Bos grunniens*).

144 "cartazonus": Aelian would seem to be describing some cross between a rhinoceros, unicorn, and yak. In any event, the creature at issue here has not been identified. In *India and the Hellenistic World* (Finnish Oriental Society, 1997), Klaus Karttunen attributes the story to Ctesias, though other scholars attribute it to Megasthenes. See also Chris Lavers, *A Natural History of Unicorns* (Granta Books, 2009).

145 See also page 118 and its note. Croton, a power on the Ionian coast of southern Italy, crushed Sybaris, whose name was a byword for luxuriousness in antiquity and remains so today, in war in 510 BCE. Some scholars consider the Sybarite horses forerunners of the famed Lipizzaner stallions of Slovenia. The bloodline is probably not identical, but certainly their dancing ability is similar.

146 "The Macedonians coated some pigs with pitch": This story was famed in antiquity. See John M. Kistler, *War Elephants* (Greenwood, 2006), 90–92.

148 "Cleitarchus says": Several python species found in India, in fact, grow to even larger size, so this is not a far-fetched claim. The second snake is possibly the slender coral snake (*Callophis melanurus*). See K. G. Charpurey, *The Snakes of India* (Fabri Press, 2008), 59–61.

The country of the Troglodytes is roughly the present-day northern Red Sea region. The largest viper known is about a quarter the size of the one Aelian describes here. A fossil snake found in Egypt, *Gigantophis*, measured about fifty feet, making it the largest of all the known snakes—but it was extinct long before humans came along.

149 "sea monsters": European eels grow to great lengths throughout the Mediterranean. See Sandra Casellato, "European Eel: A History Which Must Be Rewritten," *Italian Journal of Zoology* 69, no. 4 (2002): 321–324.

In Aelian's time, a trireme would have been about a hundred feet long, big enough to accommodate a crew of two hundred sailors. Gedrosia corresponds to coastal Pakistan.

150 Aelian seems to be describing a primate, possibly the gorilla or chimpanzee. In medieval bestiaries, the onocentaur was an obvious symbol of male lust.

151 "Dinon says": Dinon of Colophon (ca. 350 BCE) was a Greek historian. Aelian may be describing, respectively, the Abyssinian ground hornbill, warthog, and Ethiopian highland sheep.

Toads of the *Bufo* genus produce a powerful neurotoxin, an alkaloid called bufotenin. It cannot be introduced into humans by a mere gaze, however.

The Pillars of Hercules are the Straits of Gibraltar. For Eudoxus, see page 53 and its note.

152 The Veneti are commemorated in the name of present-day Venice, which is better known for pigeons than jackdaws.

For Amyntas, see page 53 and its note.

153 According to Strabo, Nebuchadnezzar built Teredon, a city on the Persian Gulf about two hundred miles from Babylon. Megasthenes adds that Nebuchadnezzar established several hanging gardens there.

Galatians: The Galatians were Celts from Thrace who settled in the central highlands of what is now Turkey. They are best known today as the recipients of a chiding letter from Saint Paul.

153–54 "oxyrhynchi": That is, sturgeons, though ancient sources also give the name to the elephantfish of Africa. Sturgeon glue is widely used to this day, particularly in the restoration of ancient artwork. See Tatyana Petukhova and Stephen D. Bonadies, "Sturgeon Glue for Painting Consolidation in Russia," *Journal of the American Institute for Conservation* 32, no. 1 (Spring 1993): 23–31. Located in west-central Iran, Ecbatana was a capital of several generations of Persian kings, among them Cyrus the Great.

154 "Herodotus tells us": 7.125.

155 Astaboras River: The Astaboras, identified by Strabo, is the Atbarah River, which rises in northwestern Ethiopia and flows into the Nile River in northern Sudan. Aelian's geography is confused. Lake Aoratia is Lake Tana.

Gregory McNamee lives on a small ranch in southern Arizona that has been populated over the years by wolves, hawks, coyotes, hounds, phainopeplas, javelinas, hummingbirds, wildcats, cardinals, a dazzling variety of snakes, and even camels. He has written and edited many books about animals and nature, among other topics, including *A Desert Bestiary*, *A Desert Reader*, and *Otero Mesa: Preserving America's Wildest Grassland*. He holds degrees in classics and English and has also published book-length translations of Aesop and Sophocles. McNamee is a contributing editor and consultant in world geography to the *Encyclopaedia Britannica* and a regular contributor to its weblog, as well as a columnist for its *Advocacy for Animals* pages.